XIX Conference on Plasma Surface Interactions 2016

Journal of Physics: Conference Series Volume 748

Moscow, Russia
28 – 29 January 2016

Editors:

V.A. Kurnaev
Yu M. Gasparyan

ISBN: 978-1-5108-3059-2
ISSN: 1742-6588

Printed from e-media with permission by:

Curran Associates, Inc.
57 Morehouse Lane
Red Hook, NY 12571

Some format issues inherent in the e-media version may also appear in this print version.

Printed by Curran Associates, Inc. (2016)

For permission requests, please contact the Institute of Physics
at the address below.

Institute of Physics
Dirac House, Temple Back
Bristol BS1 6BE UK

Phone: 44 1 17 929 7481
Fax: 44 1 17 920 0979

techtracking@iop.org

Additional copies of this publication are available from:

Curran Associates, Inc.
57 Morehouse Lane
Red Hook, NY 12571 USA
Phone: 845-758-0400
Fax: 845-758-2633
Email: curran@proceedings.com
Web: www.proceedings.com

Table of contents

Volume 748

XIX Conference on Plasma Surface Interactions
28–29 January 2016, Moscow, Russia

Accepted papers received: 1 September 2016
Published online: 23 September 2016

Preface

Papers

Journal of Physics: Conference Series **748** (2016) 011001 doi:10.1088/1742-6596/748/1/011001

Preface

V A Kurnaev, Yu M Gasparyan

National Research Nuclear University MEPhI (Moscow Engineering Physics Institute), 31 Kashirskoe highway, Moscow 115409, Russia

Corresponding authors e-mail: kurnaev@plasma.mephi.ru, YMGasparyan@mephi.ru

The 19th Conference on Plasma Surface Interactions (PSI-2016) was held at National Research Nuclear University MEPhI (Moscow Engineering Physics Institute) on 28-29 January 2016. This is a traditional annual meeting organized by MEPhI for many years and devoted to the recent achievements of Russian scientists and foreign colleagues in experimental and computer simulations of plasma and its components' interactions with plasma faced materials (PFM) for fusion devices and plasma technologies. The conference agenda covered a broad list of topics including:

- plasma induced erosion and modification of materials;
- radiation damage in materials;
- lithium and liquid metals as PFM;
- modeling of plasma surface interaction and processes in SOL of tokamak plasma;
- diagnostics of plasma-surface interaction.

The aim of the conference was to present and discuss new results in PSI in a wide audience with different areas of expertise. The important feature of the conference is to give PhD and undergraduate students the opportunity for approbation of results of their scientific activity and improve their knowledge in the novel directions of R&D in the field of fusion and plasma technologies. In total, more than 80 experienced and young researchers participated in the conference.

These Proceedings contain 21 papers selected for publication, which were reviewed by the invited international team of editors (T. Tanabe, S. Petrovich, Ch. Grisolia, Yu. Martynenko, S. Krasheninnikov, L. Begrambekov, A. Pisarev).

The Conference as well as editing of this issue were supported by National Research Nuclear University MEPhI in the framework of the Russian Academic Excellence Project (contract No. 02.a03.21.0005, 27.08.2013).

Finally, we would like to thank all of the speakers, participants and organizing committee members for their contribution to the conference.

Acknowledgements

The organizers of the conference would like to thank for the support from National Research Nuclear University MEPhI in the framework of the Russian Academic Excellence Project (contract No. 02.a03.21.0005, 27.08.2013).

Journal of Physics: Conference Series **748** (2016) 011002 doi:10.1088/1742-6596/748/1/011002

Peer review statement

All papers published in this volume of *Journal of Physics: Conference Series* have been peer reviewed through processes administered by the proceedings Editors. Reviews were conducted by expert referees to the professional and scientific standards expected of a proceedings journal published by IOP Publishing.

Metals under condition relevant to transient events in ITER

Yu V Martynenko

NRC «Kurchatov Institute», 1 Kurchatov sq., Moscow 123098, Russia
National Research Nuclear University MEPhI (Moscow Engineering Physics Institute), 31 Kashirskoe highway, Moscow 115409, Russia

Corresponding author e-mail: martyn907@yandex.ru

Abstract. Droplet erosion and shielding plasma layer are shown to be closely connected and should be considered together self consisted. Droplet erosion is caused by plasma flow over the melt metal surface. Shielding plasma is produced from droplets evaporation. Experimental data analysis and theoretical models for these processes are presented.

1. Introduction

Problem of divertor and first wall erosion is a key problem for ITER creation. These tokamak components are loaded by the most intensive plasma heat action at transient phenomena such as plasma disruption and ELM. Heat load on ITER divertor at ELM is expected to be $Q = 0.2 - 5$ MJ/m^2 at pulse duration $\tau = 0.1 - 1$ ms, and at plasma disruption to be $Q = 10 - 100$ MJ/m^2 at $\tau = 1 - 10$ ms [1, 2]. Plasma accelerator QSPU has plasma flow with parameters like those expected in ITER at disruption and ELMs [3]. However, pressure of QSPU plasma flow is several atmospheres, whereas pressure of plasma flow at ELM is expected to be $10^2 \div 10^3$ Pa and at disruptions to be $10^2 \div 4{\cdot}10^3$ Pa. The most destructive damages for divertor is breaking, studied in [4], as well as the melt layer fast transport from one place to the other [5,6]. The latter phenomenon results in the most intensive thinning of plasma facing material. Droplet erosion at temperature below boiling temperature is the main mechanism of metal mass loss.

2. Metals erosion data analysis

The works [7,8] show that energy absorbed by target reaches saturation when power density of the initial plasma flow increases, provided absorbed energy density at saturation for tungsten and graphite equals approximately the same value $Q_{abs} = 0.4 \div 0.5$ MJ/m^2. However at power density of the initial plasma flow corresponding to adsorbed energy saturation, mass losses of metals start to grow [9]. Evaporation from the surface, considered as the main mechanism of mass loss [10], is not high because movement of melt metal layer does not permit temperature to rise above the melting temperature (see Table I). Droplets seen during the pulses [9] allow us to propose that droplet erosion is the main reason of mass loss. However, emission of 100 droplets per pulse with average size ~30 μm from target 6 cm diameter at Q = 1.6 MJ/m^2 [9] corresponds to the specific mass loss $\Delta m \approx 2{\cdot}10^{-2}$ g/m^2, whereas measured mass loss at Q = 1.6 MJ/m^2 was $\Delta m \approx 5$ g/m^2 per pulse. In earlier work [12] on plasma accelerator MKT with short pulse 60 μs, but high energy density 300 kJ/m^2 much more intensive droplets emission was observed (see figure 1). The droplets size emitted at MKT pulse was ~ 1 μm and below. Estimate shows that mass loss caused by small droplets emission is much more than that caused by large droplets (~30 μm) and can explain observed mass loss. In work [6] two mechanisms of droplet emission were pointed. Both mechanisms are sequences of the plasma flow over melt metal surface, which leads to Kelvin-Helmholtz instability and waves on the surface.

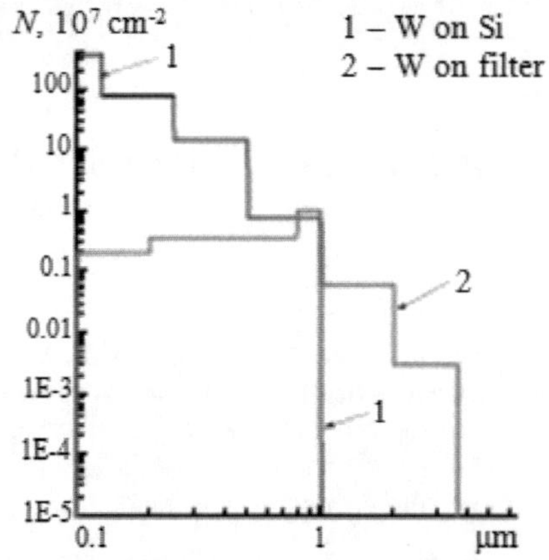

Figure 1. Size distribution of tungsten droplets at exposition in plasma accelerator MKT at $Q = 0.3$ MJ/m^2 $\tau = 60$ μs, 10 pulses. Droplets were collected: 1 - on Si collector capturing droplets flying parallel to the target surface, 2 - on basalt filter capturing droplets flying normal to target surface.

The first mechanism [13] is blowing out the wave crests by the plasma wind. In the work [13] this mechanism was believed to be responsible for small ~ 1 μm droplets emission. The second mechanism is a result of moving wave instability, which leads to one wave intercepting another one resulting in droplets emission. This mechanism explains emission of droplets with size 30—60 μm.

3. Waves on the melt metal surface and droplet emission

Waves emerged due to Kelvin—Helmholtz instability suggests existence of plasma flow over the melt metal surface [14]. In this case wave length λ, frequency ω and increment γ are the following

$$\lambda = 3\pi\alpha/2P \tag{1}$$

$$\omega = (2^{5/2}/3)\cdot(P^{3/2}/\alpha\rho'^{1/2})(\rho'/\rho)^{1/2}, = 1.9\cdot(P^{3/2}/\alpha\rho'^{1/2})(\rho'/\rho)^{1/2} \tag{2}$$

$$\gamma = (2^{5/2}/3^{3/2})\, P^{3/2}/\alpha\rho'^{1/2} \approx 1.09\cdot P^{3/2}/\alpha\rho'^{1/2}, \tag{3}$$

where α is the surface tension, ρ' and U are the density and the velocity of plasma flow, respectively, ρ is the metal density, $P = \rho' U^2/2$ is the pressure of plasma flow. It is noticeable that $\omega << \gamma$ by $(3\rho/\rho')^{1/2} \approx 10^3$ times, in the most cases $\omega \sim 1/\tau$ and motion of waves is not possible. At pulse duration τ waves can arise only if $\gamma > 1/\tau$. This condition means that $P > P_{cr} = 0.94\cdot\alpha^{2/3}\cdot\rho'^{1/3}\cdot\tau^{-2/3}$. At $\tau = 1$ms the pressure of plasma flow should be $P > 7\cdot10^3$ Pa for tungsten, ($\lambda_{max} = 0.15$ mm), and $P > 2.2\cdot10^3$ Pa for beryllium, ($\lambda_{max} = 3$ mm).

Let us consider droplets emission as a result of the wave crests blowing out. The velocity of plasma flow over the waves is much higher than the velocity of wave motion. The plasma wind acts on the wave crest. Due to the viscosity the wave crest with height Δ can be considered as liquid on the solid substrate during the time t

$$t = \Delta^2/\nu \tag{4}$$

where ν is the kinematic viscosity. If this wave crest is shifted for the time t on the distance equal the length of wave crest base d, this wave crest will be taken away as droplets. If the plasma wind pressure is balanced by surface tension [6], the length of wave crest base d equals

$$d = (\Delta \cdot r)^{1/2},\ r = \alpha/P \tag{5}$$

The time of wave crest shift on the distance d is

$$t = (2d/a)^{1/2},\ a = P\Delta/\rho\Delta d = P/\rho d \tag{6}$$

where a is the wave crest acceleration. The wave crest height can be found from (4), (5) and (6),

$$\Delta = (2\alpha\rho\nu^2)^{1/3}P^{-2/3} \tag{7}$$

It is noticeable that velocity of the wave crest reaches the value $v = P\Delta^2/\rho\nu d$, at which the plasma wind force equals to the friction force, just for the time of the crest shift. This means that the wave crest shift occurs in the regime of acceleration. Blowing away of one wave crest occurs after the previous wave crest is blown away in the time t. The wave height is limited by the crests removing and equals to $H = \Delta/\gamma_t$. The crest is removed from the whole front of the wave, and then the crest is divided into droplets with the volume ~ $(\Delta d)^{3/2}$ each. So blowing away of the crest from the wave with radius R results in appearance of $\pi R/(\Delta d)^{1/2}$ droplets with size

$$r = (\Delta d)^{1/2} = (2\rho)^{1/4}(\alpha\nu)^{1/2}P^{-3/4} \tag{8}$$

Droplets size decreases with the plasma wind pressure increase P. Droplets emitted in the region with maximal P have the smallest size. Droplets at the edge of plasma flow region where $P \approx P_{cr}$ have the maximal size. Table I shows that blowing away of the wave crests also results in large droplets emission. Number of droplets emitted from an area unit per pulse is

$$N_{dr}(P) = (\tau/t)(1/\lambda r) = \tau(2^{1/12}/3\pi)(\nu^{5/6}\alpha^{13/6}\rho^{11/12})^{-1}\cdot P^{37/12} \tag{9}$$

If the pressure distribution is $P \sim P_0 \cdot \exp(-R^2/\sigma^2)$, one can find the size distribution $N_{dr}(r)$ of droplets

$$N_{dr}(r) = \mathrm{d}N_{dr}/\mathrm{d}R \cdot 2\pi R \cdot \mathrm{d}R/\mathrm{d}r \approx \tau\, \nu^{23/18}\cdot\alpha^{-1/9}\cdot\rho^{1/9}\cdot P_0^{\ 37/24}\cdot\sigma^2\cdot r^{-46/9} \tag{10}$$

Mass loss is determined by the small droplets emission and mass loss per pulse is

$$\Delta m = \rho\tau\int r^3\cdot(\mathrm{d}N_{dr}/\mathrm{d}r)\cdot \mathrm{d}r \sim P_0^{\ 25/8} \tag{11}$$

Mass loss grows steeply with the growth of the plasma flow pressure in accordance with [5]. The table I gives characteristics of droplets emission and mass loss for tungsten and beryllium at the initial plasma flow pressure corresponding to the energy density flow exceeding twice the threshold energy density for melting.

Table 1. The number of droplets emitted per QSPU pulse with $\tau = 0.5$ ms (N); the mass loss per pulse calculated with formula (11) (Δm), the mass loss caused by evaporation from the surface (Δm_{vapor}) and the mass loss measured experimentally [5] (Δm_{exp}), the maximal size of droplets emitted at wave crest blowing away (r_{max}).

	N, m^{-2}	Δm, g/m^2	Δm_{vapor}, g/m^2	Δm_{exp}, g/m^2,	r_{max}, µm
W, P_0= 3 atm	$5\cdot10^{12}$	20	$6\cdot10^{-2}$	3÷9	30
Be, P_0= 2 atm	$4\cdot10^{14}$	10	$(4.5÷7)\cdot10^{-3}$	0.1÷0.35 *($P \approx 1$ atm)*	37

4. Near surface layer of shielding plasma

Instability development and waves on the melt metal surface are caused by the near surface plasma layer. The thickness of this layer is higher than the wave length, but the parameters of this layer differ from the parameters of the shielding layer at ~ 1 cm from the metal surface [8]. Therefore we believe that the thickness of the near surface plasma layer is L ~ 1 mm.

The saturation of the energy absorbed by target means that the near surface plasma layer is optically thick. In this case temperature T of the near surface plasma layer can by calculated as $T = (W_{abs}/\sigma)^{1/4}$, where σ is the Stephan-Boltzmann coefficient, $W_{abs} = Q_{abs}/\tau$ is the absorbed power at saturation. At $W_{abs} \approx 1\mathrm{GW/m^2}$ [8], $T \approx 1$ eV. The energy of the particle flow to the surface is much less than the radiation energy. We will believe that the temperature of the near surface plasma layer is constant $T \approx 1$ eV. The main mechanism of the target mass loss is droplet erosion, and we believe that droplets evaporation in the near surface plasma layer is the main source of the vapor for shielding plasma. The estimate shows that tungsten droplet flying from the surface evaporates if its size is $r < 1$ μm. The larger droplets decrease the size by ~ μm. At the constant temperature of the near surface layer the pressure in this layer is determined by the plasma density. The density of the near surface layer can be calculated from the condition of the flow continuity. At normal extension of the shielding plasma the density of the near surface plasma layer is $n = q/v_{exp}$, where $q = \Delta m/M\tau$ is the flow of vapor caused by droplets evaporation, M is the metal atom mass. Small droplets contribute the most to mass loss. Therefore we take Δm from the experimental mass loss. The velocity of plasma extension is $v_{exp} = [v_T^2 - P_{ext}/\rho']^{1/2}$, were v_T is the heat velocity, P_{ext} is the external pressure. Than the pressure in the near surface plasma layer is

$$P = P_{ext}/2 + [(P_{ext}^2/4) + (Mqv_T)^2]^{1/2} \quad (11)$$

The formula (11) shows that at $q > P_{cr}/2Mv_T$, even at P_{ext}=0, the pressure in the near surface layer is sufficient for tangential plasma flow, which can cause waves to appear and droplets to erode.

There are some examples when the instability of the melt metal layer and their sequences were observed at the pressure of initial plasma flow below critical: (i) droplets emission observed in [11] (figure 1) occurs at the pressure of initial plasma flow lower than required for development of the wave structure and droplet erosion. The pressure of MKT plasma flow was 10^3 Pa, whereas critical pressure for Kelvin-Helmholtz instability on tungsten is $P_{cr} = 7\cdot10^3$Pa, (ii) other evidence of the shielding plasma influence on droplet emission is time dependence of droplets ejection obtained in [5]. The main droplets emission occurs after QSPA pulse completion. Pulse duration was 0.5 ms, whereas maximal droplet emission was observed at 1 ms and droplet emission lasted up to 2 ms. This means that droplet emission was initiated by the shielding plasma which regenerates itself by droplets evaporation, (iii) the best evidence of the shielding plasma action is simulation of the disruption mitigation on QSPA [4]. In this case, QSPA plasma flow acts on Ar gas target and Ar radiation acts on stainless steel. Radiation has zero pressure but the observed wave structure is a result of shielding plasma flow over the melt metal surface.

However, development of the Kelvin-Helmholtz instability without external pressure is possible not in all cases. Calculation and experiment [11] show that erosion of beryllium is low and $q < P_{cr}/2Mv_T$. Experiments on plasma accelerator QSPU-Be in Bochvar Research Institute of Inorganic Materials [11] show that the wave relief and the melt metal movement on Beryllium is observed only after direct plasma action on the Be target. After radiation pulses with the same energy the wave structure was absent. Instability development at the absent external pressure is possible at the normal initial plasma flow and normal shielding plasma extension. At the sliding initial plasma flow and at magnetic field the shielding plasma expansion occurs otherwise. These cases are to be considered specially.

Acknowledgements
This work was supported in part by National Research Nuclear University MEPhI in the framework of the Russian Academic Excellence Project (contract No. 02.a03.21.0005, 27.08.2013).

References
[1] ITER Physics Basis 1999 *Nucl. Fusion* **39** 2137

[2] Federici G, Skinner C H, Brooks JN et al. 2001 *Nucl. Fusion* **41** 1967
[3] Würz H, Arkhipov N I, Bakhtin V P, Safronov V M et al. 1995 *J. Nucl. Mat.* **220—222** 1066
[4] Budaev V P, MartynenkoYu V, Khimchenko L N, Zhitlukhin A M, Klimov N S, Pitts R A, Linke I, Belova N E, Karpov A V, Kovalenko D V, Podkovyrov V L and Yaroshevskay A D 2013 *Plasma Phys. Reports* **39** 910
[5] Poznyak I M, Klimov N S, Podkovyrov V L,Safronov V M, Zhytlukhin A M and Kovalenko D V 2012 *PAST Ser. Thermonuclear Fusion* **4** 23
[6] MartynenkoYu V 2014 *PAST Ser. Thermonuclear Fusion* **37** (2) 53
[7] Safronov V M, Arkhipov N I, Landman I S, Pestchanyi S E, Toporkov D A and Zhitlukhin A M 2009 *J. Nucl. Mat.* **386–388** 744
[8] Poznyak I M, Arkhipov N I, KarelovS V, Safronov V M and Toporkov D A 2015 *International Summer School on the Physics of Plasma-Surface Interaction* Moscow, MEPhI
[9] Klimov N S, Podkovyrov V L, Zhitlukhin A M, Safronov V M, Kovalenko D V, Moskacheva A A and Posnyak I M 2009 *PAST Ser. Thermonuclear Fusion* **2** 52
[10] Krasheninnikov S I, Pigarov A Yu and Lee Wonjae 2015 *Plasma Phys. Control. Fusion* **57** 044009
[11] Safronov V M *Private communication*
[12] Guseva M I, Gureev V M, Domantovskii A G, Martynenko Yu V, Moskovkin P G, Stolyarova V G, Strunnikov V M, Plyashkevich V and Vasilev V I 2002 *J. Techn. Phys.***72** 7 841
[13] Martynenko Yu V and Moskovkin P G 2000 *PAST Ser. Thermonuclear Fusion* **1** 65
[14] Landau L D and Lifshits E M 1988 *Hydrodynamics* **6** (Moscow: Nauka)

Surface biasing influence on the physical sputtering in fusion devices

I Borodkina[1,2], D Borodin[2], S Brezinsek[2], I V Tsvetkov[1], V A Kurnaev[1], C C Klepper[3], A Lasa[3], A Kreter[2] and JET Contributors*

EUROfusion Consortium, JET, Culham Science Centre, Abingdon, OX14 3DB, UK
[1]National Research Nuclear University MEPhI (Moscow Engineering Physics Institute), 31 Kashirskoe highway, Moscow 115409, Russia
[2]Forschungszentrum Jülich GmbH, Institut für Energie- und Klimaforschung - Plasmaphysik, Jülich 52425, Germany
[3]Oak Ridge National Laboratory, Oak Ridge, TN 37831-6169, USA
*See F. Romanelli et al., Proc. of the 25th IAEA Fusion Energy Conference 2014, Saint Petersburg, Russia

Corresponding author e-mail: ieborodkina@mephi.ru

Abstract. A new simplified analytical expression for the electromagnetic field in the Debye sheath in the presence of an oblique magnetic field including surface biasing effect is suggested. It is in good agreement with the numerical solution of the integral equation for the potential distribution in the Debye sheath. The energy and angular impact distributions and corresponding surface sputtering yields were analyzed in the presence of an oblique magnetic field and surface biasing. The analytical expression was used to estimate a) the effective sputtering yield of the W target with a varying negative voltage against plasma in PSI-2 linear device and b) erosion of the JET outer wall Be limiter near the ICRH antenna enhanced during RF emission.

1. Introduction

Understanding plasma-surface interaction (PSI) processes is important for the successful realization of the ITER project. PSI processes such as erosion and deposition lead to reduction of life time of plasma-facing components (PFC) in fusion devices. For correct calculation of the sputtering yields for PFCs in the presence of an oblique magnetic field the accurate expression for the sheath electric field must be included. Furthermore, in a number of experiments surface biasing (SB) up to several hundred Volts is applied to the PFC (e.g. to study the effect of ion impact energy influence), which also influences greatly the sheath potential distribution.

In the present paper the simplified analytical expression (AE) for the electromagnetic field suggested in [1] is updated to include the SB effect and an oblique magnetic field. Using this expression we reveal the influence of SB on the ion impact energy and angle distributions, which have strong influence on local physical sputtering yields.

The AE mentioned above was used to estimate the sputtering yield of the W target under Ar plasma exposure in PSI-2 linear plasma device [2]. In this experiment, the magnetic field is directed normal to the target surface. However, the additional negative SB of up to several hundred Volts (applied to the

Journal of Physics: Conference Series **748** (2016) 012002 doi:10.1088/1742-6596/748/1/012002

target to vary the sputtering ion energies) has a considerable influence on the sheath potential, the energy distributions, and thus, it must be treated as an additional parameter in the analytical solution.

In the case of modeling of PSI processes at the JET ITER-like wall (ILW), an oblique magnetic field is the most common scenario, and thus should be taken into account. The strong effect of the magnetic field angle in the energy and angle distributions highlights the relevance of the present effort in a correct estimate of sputtering yields. The suggested AE has been applied for modeling of radio-frequency (RF)-sheath enhanced erosion of the outer wall JET Be limiter near the Ion Cyclotron Resonance Heating (ICRH) antenna, aiming to improve the modeling presented in [3]. The comparison of the simulated RF-enhanced Be emission with experimental observations is presented.

2. Surface biasing influence on potential distribution in the sheath

The expression for the dependence of target potential on the SB was derived from the charge conservation law (similar to [4] (2.65)) where we take into account the difference in areas of target and return (surrounding) surfaces:

$$S_{rs} \cdot \Gamma_{rs}^{e} + S_{t} \cdot \Gamma_{t}^{e} = n_{se} c_{s} (S_{t} + S_{rs}) \tag{1}$$

where S_t, S_{rs} are the areas of the target and the return surface, Γ_t^e and Γ_{rs}^e are electron flux densities on the target and the return surfaces, n_{se} is plasma density at the sheath edge, c_s is the sound speed and $\Gamma^i = n_{se} c_s$ is the ion flux density on the surfaces. The electron flux densities reaching the target and return surfaces according to [4] are:

$$\Gamma_t^e = \frac{1}{4} n_{se} \bar{c}_e \exp\left[\frac{e(U_t - U_{pl})}{kT_e}\right] \tag{2}$$

$$\Gamma_{rs}^e = \frac{1}{4} n_{se} \bar{c}_e \exp\left(\frac{e(U_{rs} - U_{pl})}{kT_e}\right) \tag{3}$$

The ion flux density on the surfaces Γ^i is the same as without SB:

$$n_{se} c_s = \frac{1}{4} n_{se} \bar{c}_e \exp\left(\frac{eU_{sf}}{kT_e}\right) \tag{4}$$

where $\bar{c}_e$ is the thermal electron speed, U_{pl} is the plasma potential at the sheath/presheath boundary, U_{sf} is the surface floating potential relative to U_{pl} ([4] (2.60)), U_t is the target potential, U_{rs} is return surface potential.

Then, rewriting equation (1) in the following way,

$$\Gamma_t^e = \frac{n_{se} c_s}{\dfrac{S_t}{S_t + S_{rs}} + \dfrac{S_{rs}}{S_t + S_{rs}} \cdot \dfrac{\Gamma_{rs}^e}{\Gamma_t^e}} \tag{5}$$

we substitute equations (2)-(4) into (5) and obtain the expression for the dependence of target potential on the SB:

$$\frac{e\left(U_t - U_{pl}\right)}{kT_e} = \frac{e\left(U_{sf}\right)}{kT_e} - \ln\left[\frac{S_t}{S_t + S_{rs}} + \frac{S_{rs}}{S_t + S_{rs}} \exp\left(\frac{e\Delta U_{bias}}{kT_e}\right)\right] \tag{6}$$

where $\Delta U_{bias} = U_{rs} - U_t$ is a negative surface biasing.

As the area of the return surface is usually much larger than that of the target ($S_t \ll S_{rs}$), the target potential in the presence of SB can be calculated as:

$$\frac{e(U_t - U_{pl})}{k \cdot T_e} = \frac{e(U_{sf})}{kT_e} - \frac{e\Delta U_{bias}}{kT_e} \tag{7}$$

Further, PSI processes are determined to a large extent by the Debye sheath (DS) and magnetic pre-sheath (mps) effects, where the strong electric field is located. As the SB influences the potential distribution only in the DS [5], the analytical potential expression obtained in [1] can also be used here for the mps:

$$\psi(\xi) = \psi_{mps} \cdot \exp(-2 \cdot (\xi - \xi_{mps}) / L_{mps}) \tag{8}$$

where $\psi = e \cdot (U - U_{pl}) / k \cdot T_e$ is the normalized potential, ξ is the distance from the surface in units of the Debye length, $\psi_{mps} = \ln \cos \alpha$ is the normalized potential drop in the mps as derived in [4], α is an angle between the magnetic field and the surface normal, $L_{mps} = k \cdot \rho_{iCs} \cdot \sin \alpha$ is the length of the magnetic pre-sheath, the factor k can be chosen between 2 – 3 [1], ρ_{iCs} is the Larmor radius corresponding to the ion acoustic velocity in units of the Debye length, ξ_{mps} is the distance to the mps/DS boundary.

For the potential distribution in the DS, the approximation of the numerical solution to the Poisson equation derived in [1] with the boundary conditions at the surface (equation (7)) and at the DS/mps (equation (8)) can be taken as the following:

$$\psi(\xi) = \psi_w + Q - Q \cdot \exp(-a \cdot \xi - c \cdot \xi^2) \tag{9}$$

where ψ_w is the dimensionless target potential (equation (7)), parameters a and Q are derived in [1], the parameter c and ξ_{mps} were determined by numeric iterations. The resulting sheath electric field potential is in good agreement with the numerical solution of the integral Poisson equation in the DS. Figure 1 presents the influence of the SB on the potential profiles calculated using the suggested analytical expression given by equations (8) and (9). The potential changes so that both the potential drop and the length of the DS increase.

The suggested model can be easily applied for various experiment conditions. Figure 2 illustrates the influence of the SB voltage on the angular distributions of the impinging D^+ ions calculated using the suggested AE for different magnetic field intensities and angles. Two opposite trends in the shift of the mean ion impact angle with SB are revealed: the main trend is the decrease of the mean impact angle with biasing, although under strong magnetic fields and at grazing angles ($\xi_{mps} > \rho_{iCs} \cdot \sin \alpha$) the electric field influence on the ion impact angle becomes inessential and the mean angle increases.

Further, the influence of SB on the effective sputtering coefficient of Be by incident D ions was calculated by integrating the Eckstein formula [6] for the angle and energy distributions obtained above. In most cases the effective sputtering yield increases by a factor of 3 with increasing the SB up to 200 V. However, at high magnetic field intensities at grazing angles the yield rises up to 10 times due to the angular factor.

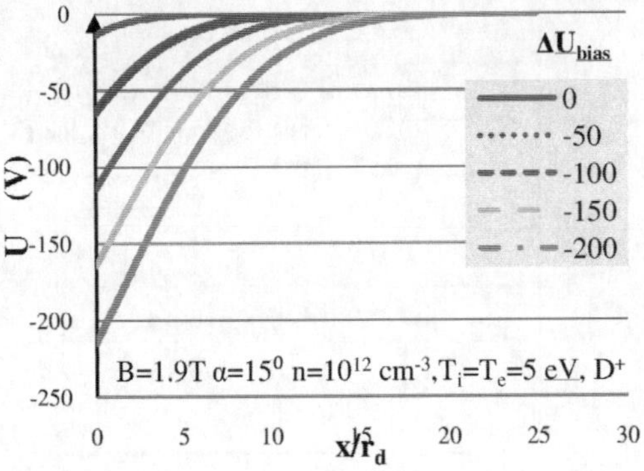

Figure 1. Potential profiles obtained using analytic expression (equations (8), (9)) for different biasing voltages.

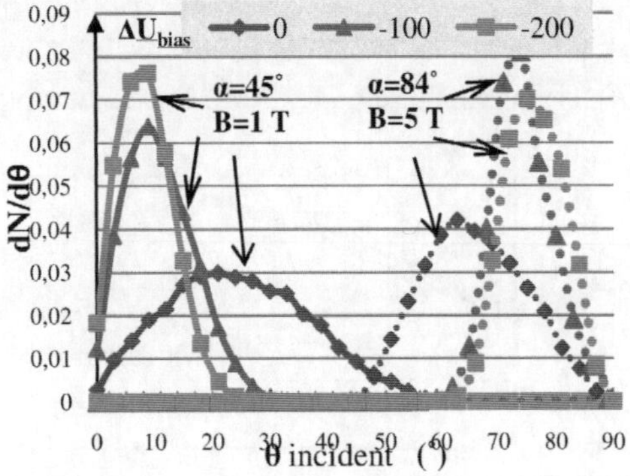

Figure 2. Surface biasing influence on the angular distributions of incident D ions for different values and angles of magnetic field to the surface (n=10^{12} cm^{-3}, T_i=T_e=5 eV).

3. Surface biasing influence on W target sputtering in the linear device PSI-2

The AE mentioned above was applied to the study of sputtering of the W target under Ar plasma exposure in PSI-2 linear plasma device [2]. Experimentally, the neutral W (400.9 nm) radiation intensity profiles were measured by a spectrometer along the normal distance to the target for various target voltages.

Using the expression (9) we calculated the energy and angular ion distributions and corresponding effective sputtering yield for each target voltage. In figure 3 the calculated data, normalised to the sputtering coefficient at $|U_t - U_{pl}|$ = 100V, are compared to the respective experimental data. The experimental values of the effective sputtering yield for each target voltage were obtained by integrating the W intensity profiles. The good agreement of simulated and experimental results proves the applicability of the suggested AE. The results obtained with the AE improve the estimates calculated using simplified sheath assumptions up to 10% in this particular case. However, the AE application efficiency increases with the ion temperature and magnetic field angle.

4. Simulation of enhanced by RF-emission erosion of JET Be limiter

The AE derived above has also been applied for modelling of 2-3 times enhanced erosion of the outer wall JET Be limiter modulated by toggling of ICRH RF-antenna ‘C’ [3]. This effect has been associated with magnetic connections between the affected PFCs and high electric fields induced by the antenna, the latter being represented in modeling by an additional negative SB of up to 200 V [7]. Figure 4 presents the sputtering coefficients in the assumption of the low-recycling plasma scenario calculated with the AE and obtained in the earlier simulations which did not account for the influence of the oblique magnetic field [3]. These sputtering coefficients were calculated assuming 50% D concentration in the surface interaction layer (‘ERO-min’) [8]. For comparison the case of a pure beryllium target (‘ERO-max’) was also calculated with the AE. It is shown that for ‘ERO-min’ fit, the calculated sputtering yield increases 2-3 times with biasing, which is in qualitative agreement with the experimental results (α=85.8°, B=1.9 T, n=10^{12} cm^{-3}, T_i=T_e=5eV). This provides additional confidence in ‘ERO-min’ fit for the physical sputtering yields for the plasma-wetted areas of PFCs. The properly treated angular factor in the sputtering yield leads to reproducing of the experimental erosion increase mentioned above.

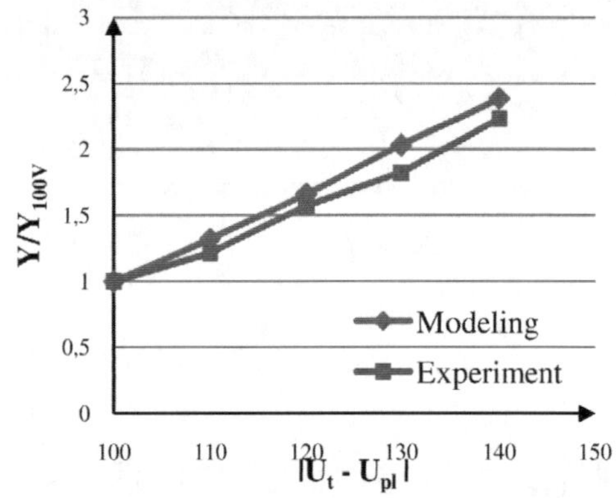

Figure 3. Comparison of the experimental and simulated dependences of the sputtering yields of W target by Ar ions impact on ion incident energy in the linear device PSI-2 (α=0°, B=0.1T, n = 0.45·10^{12} cm^{-3}, T_i = 1eV, T_e = 4eV, U_{pl} = -10V, Ar^+).

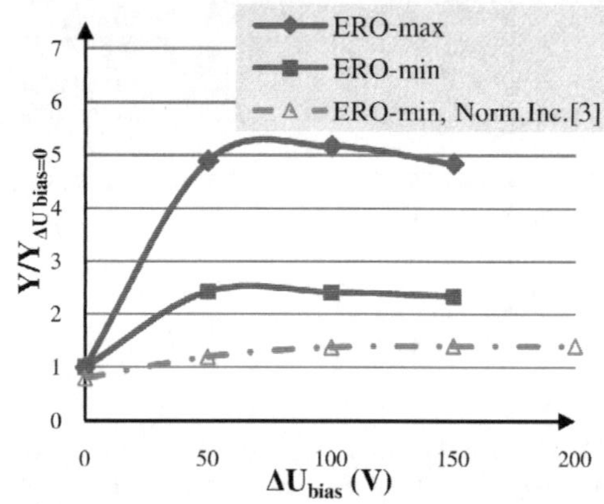

Figure 4. Comparison of the simulation with AE for different surface content, the earlier ERO modeling (ERO-min for normal incidence) and experimental observations (rectangle) of Be limiter erosion.

5. Conclusion

The simplified analytical expression for the electromagnetic field in the DS suggested in [1] is now updated to include surface biasing effect. The energy and angular impact distributions and surface sputtering yields were computed in the presence of an oblique magnetic field and surface biasing. Two opposite trends of the mean ion impact angle shift with surface biasing are revealed: in most cases, the mean impact angle decreases with biasing, however, under strong magnetic fields and at grazing angles the mean angle increases.

The AE was used to estimate the sputtering yield of the tungsten target under Ar plasma for various target voltages in PSI-2 experiments. The simulation and experimental results are in a good agreement. The same AE has been applied for improving earlier estimates [3] of enhanced by RF-emission erosion of the outer wall JET Be limiter near the ICRH antenna. It is shown that an additional negative surface biasing more than 50 V can explain the observed 2-3 fold increase in erosion (characterized by Be spectroscopy), assuming 50% D concentration in the surface interaction layer. The updated model, however, leads to an increased effect due to the synergy with the properly treated angular factor in the sputtering yield. This leads to a qualitative change in the results: for instance with the earlier model [3] the increase in erosion observed experimentally could not be reproduced under the low-recycling plasma assumption. In addition, the roughness of the substrates can influence the surface sputtering as is shown in [9]. Therefore, some of our earlier modeling results and conclusions for ERO modeling of enhanced by RF-emission erosion of the outer wall JET Be limiter near the ICRH antenna [5] should be re-visited.

Acknowledgements

This work has been initiated within the MEPhi – FZJ strategic partnership. The views and opinions expressed herein do not necessarily reflect those of the European Commission. The authors (I. Borodkina, I.V. Tsvetkov and V.A. Kurnaev) also would like to thank for the support from National Research Nuclear University MEPhI in the framework of the Russian Academic Excellence Project (contract No. 02.a03.21.0005, 27.08.2013).

References

[1] Borodkina I et al. PET-2015 submitted to *CPP*
[2] Eksaeva A et al. 2015 *Proc. of the XXII Int. Conf. ISI-2015* 82
[3] Klepper C C et al. 2016 *Phys. Scr.* 014035
[4] Stangeby P 2000 *The plasma boundary of magnetic fusion devices* (Plasma physics series) (London: IOP Publishing Ltd)
[5] Weinlich M, Carlson A 1997 *Phys.Plasmas* **4 (6)** 2151
[6] Behrisch R, Eckstein W 2007 *Topics Appl. Physics* **110** 33
[7] Bobkov V V et al. 2010 *Nucl. Fusion* **50** 035004
[8] Borodin D et al. 2013 *J. Nucl. Mater.* **438** 267
[9] Shuyu D et al. 2015 *J. Nucl. Mater.* **463** 372

Boron carbide coating deposition on tungsten substrates from atomic fluxes of boron and carbon

Y Sadovskiy, L Begrambekov, A Ayrapetov, I Gretskaya, A Grunin, M Dyachenko, N Puntakov

National Research Nuclear University MEPhI (Moscow Engineering Physics Institute), 31 Kashirskoe highway, Moscow 115409, Russia

Corresponding author e-mail: yaroslav@plasma.mephi.ru

Abstract. A device used for both coating deposition and material testing is presented in the paper. By using lock chambers, sputtering targets are easily exchanged with sample holder thus allowing testing of deposited samples with high power density electron or ion beams. Boron carbide coatings were deposited on tungsten samples. Methods of increasing coating adhesion are described in the paper. 2 μm boron carbide coatings sustained 450 heating cycles from 100 to 900 C. Ion beam tests have shown satisfactory results.

1. Introduction

Currently, one of the main issues considering first wall materials of fusion devices is material erosion and dust generation [1-3]. First wall materials are being sputtered and codeposited layers are being formed. Protective boron carbide coating was proposed to prevent tungsten sputtering and consequent plasma contamination [4]. In a fusion device this type of coating could be deposited in-situ by using carborane vapors. In this case the coating would be formed of boron and carbon atoms originating from carborane molecules decomposition in plasma. For research purposes it is more convenient to deposit such coating by sputtering two separate targets made of boron and graphite and thus generating fluxes of boron and carbon atoms.

2. Boron carbide deposition on tungsten

Coating deposition as well as its consequent testing was performed on the CODMATT device (COating Deposition and MATerial Testing) [5]. Several modifications were done to the device prior to experimental program. The device could be easily transformed from deposition configuration to material testing configuration. Schemes of both configurations are presented in figure 1.

The device has one vacuum chamber (1) with a plasma chamber (2) inside. Two heated cathodes (3) and a ring shaped anode (4) on a motion feedthrough (5) are located inside the plasma chamber. The cathodes can be used independently or turned on simultaneously for higher emission. The device has two lock chambers – top (6) and bottom (7) equipped with gate valves (8) and special sealings (9) for motion feedthroughs. In deposition configuration the substrate (10) is introduced into the vacuum chamber on motion feedthrough (11) through the top lock chamber. The substrate is mounted on a copper table which could be biased in case the sample needs to be sputtered before coating deposition. The target unit consists of: a graphite cup (12), two anti-dynatron electrodes (13, 14) which connect with slip-contacts (15) when moved into correct position by motion feedthrough (16). A quadrupole mass-spectrometer with separate vacuum system is connected to the vacuum chamber through a leak valve. Thus it is possible to monitor working gas composition changes during coating deposition and sample testing.

A mix of graphite and boron powders and alcohol is prepared and applied on the surface of the graphite cup. This paste-like substance becomes solid crust when alcohol vaporizes. Graphite to boron powder ratio is 4:1 thus allowing the right stoichiometry of the coating.

Argon is used as a working gas. The device allows adding of hydrogen or deuterium to the working gas for the research of hydrogen trapping into coatings being deposited. Using of a heated cathode gives high control of the discharge current and thus plasma density. When high negative bias is applied to the target a ring-shaped ion beam is formed, exiting from the opening in the anode. Typical current and voltage values of the ion beam are 30-40 mA and 10-12 kV. Higher negative bias is applied to the anti-dynatron electrodes thus suppressing emission of secondary electrons from the target. Outer anti-dynatron electrode is made of tungsten and could be used a tungsten target in case source of tungsten atoms is needed for the deposition.

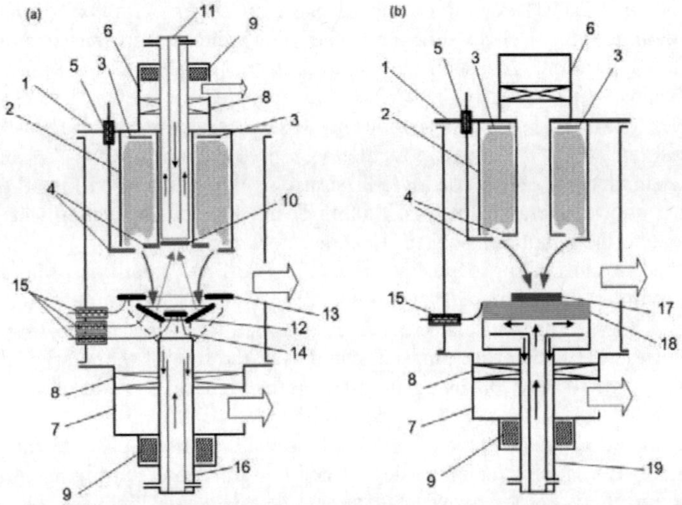

Figure 1. Scheme of the CODMATT device in two configurations: coating deposition (a), material testing (b). 1 – vacuum chamber, 2 – plasma chamber, 3 – heated cathode, 4 – anode, 5 – motion feedthrough, 6 – top lock chamber, 7 – bottom lock chamber, 8 – gate valve, 9 – motion feedthrough sealing, 10 – substrate, 11 – substrate motion feedthrough, 12 – target, 13 – outer anti-dynatron electrode, 14 – center anti-dynatron electrode, 15 – electric feedthrough, 16 – target motion feedthrough, 17 – sample being tested, 18 – copper table, 19 – testing table motion feedthrough.

Tungsten substrates of 15x15x1 mm size were used. The substrates were polished with sand paper of up to 2500 grade and then cleaned in alcohol in ultrasonic bath. Deposition regimes will be described below.

Deposited coatings were tested in various regimes with electron and ion beams. To allow this, the experimental device was reconfigured in the following way: the sample being tested (17) was fixed to a cooled copper table (18) which was introduced into the vacuum chamber through the bottom lock chamber similar to the target unit. The copper table is mounted on top of tube-shaped insulator and cooled with non-conductive silicon based fluid PMS-5 (poly-methyl-siloxane). This fluid is circulated in a closed loop with a heat exchanger which is in turn cooled by water. Using of a non conductive fluid allows direct cooling of metal parts which are biased with high voltage. The copper table was designed to dissipate high power density heat loads and transfer heat from a small spot at the center of its top surface to a ribbed bottom surface actively cooled by the cooling fluid.

By varying the distance from sample table to the anode, value of the bias voltage and discharge current the shape of the ion beam could be changed as desired. By focusing the beam to a small spot one can achieve high power densities such as 20-40 MW/m^2. Total power of the high voltage supply is 4 kW. Without active cooling sample temperatures could be higher than 2000 K. Since the high voltage power supply is fully programmable, automatic thermocycling procedures could be performed. Pulse frequency could be as high as 100 Hz with independent control over duty cycle.

The polarity of the high voltage supply could be easily changed thus allowing electron beam irradiation. Thus one can compare ion beam and electron beam effects on the sample in almost identical irradiation conditions.

3. Deposition procedure optimization and testing of the coating.

Preliminary experiments have shown that when sputtering 4:1 boron and graphite powder mix coating composition is close to B/C=4:1. Coating composition was measured with Oxford Instruments X-Act EDS analyzer. Coatings with > 3 μm thickness had poor adhesion to the substrate, flakes were observed right after deposition. The deposited samples were tested both with pulse and stationary electron and ion beams. After the tests the surface was analyzed in TESCAN VEGA 3 SEM and in Quanta 200 3D FIB SEM, where a cross section of the surface was made by focused ion beam and depth structure was analyzed. Thus it was possible to observe delamination of the coating, pores and voids. The main characteristic of the coating adhesion was its ability to withstand pulse heat loads having certain maximum temperature with no flaking or cracks. Thermocycling tests up to 300, 600 and 900 C were done to the samples deposited. Heating rate was 5 C/sec.

First samples had 2 μm coating deposited at 200 °C with no interlayer. Massive flaking was observed on these samples already after 10 cycles with 300 °C maximum temperature.

This result proves the need for special procedures intended for increasing coating adhesion to the substrate. In case of the ITER divertor, tungsten could play the role of the interlayer material. Thus, further experiments were focused on using tungsten as the interlayer material to increase coating adhesion.

The first stage included establishing of a smooth transition from substrate material to coating material. This method is common for increasing of coating adhesion. By mixing of materials in the interlayer there is no sharp border between different materials at which delamination can occur. Deposition procedure started with tungsten deposition on the substrate using outer anti-dynatron electrode as a target. Then bias voltages were varied constantly to smoothly decrease tungsten sputtering and at the same time start target sputtering. These voltages changes were preprogrammed in the high voltage power supply. Final bias values were the same as in previous experiments. Thus, an interlayer with ~200-300 nm thickness was deposited, establishing smooth transition from substrate to coating material. 2 μm coatings were deposited using this procedure. These coatings passed 100 thermocycles at 600 °C maximum temperature with no surface defects. When the maximum temperature of thermocycling was increased to 900 °C flaking was observed. Cracks in the coating often corresponded to cracks in the substrate. Crack sides were lifted off the substrate which is a sign of compressive stress in the coating. By analyzing the composition of the back side of a delaminated flake a conclusion about the layer of delamination was made. Tungsten content on the back side of the flake indicated that delamination occurred at the interlayer depth. Most probably, deposited tungsten was weakly bonded to substrate material.

Thus, at the next stage, substrate pre-sputtering was performed prior and during tungsten deposition. This serves two goals: cleaning of the substrate surface prior to deposition and mixing of substrate material with newly deposited material. The interface between substrate and deposited layer of tungsten is almost eliminated in this way. Then, the interlayer was generated as described above – smooth transition from tungsten to boron carbide. 2 μm coatings deposited in this way have sustained 450 thermocycles at 900 °C maximum temperature with no coating defects observed. Then, ion beam tests were performed with the following parameters: H^{2+} ions with 6 keV energy at 0,5 MW/m^2 energy density, pulse duration 100 sec. Sample temperature during this tests did not exceed 700 °C which is the temperature expected for ITER divertor tiles. After 120 such pulses sputtering artifacts were observed on the coating surface with no signs of flaking or cracking.

4. Conclusion

A multifunctional device allowing both boron carbide coating deposition and testing of materials and coatings with high heat and particle loads is presented.

A method of boron carbide coating deposition with good adhesion to tungsten was developed. This method could be realized in ITER conditions in-situ.

Deposited coatings sustained 450 thermocycles from 20 to 900 C with no visible surface degradation. The coatings sustained pulse H^{2+} ion irradiation with 6 keV ion energy at 0,5 MW/m^2 energy density and 700 °C maximum temperature with no cracking.

Acknowledgements

This work was supported by National Research Nuclear University MEPhI in the framework of the Russian Academic Excellence Project (contract No. 02.a03.21.0005, 27.08.2013) and by the Ministry of Education and Science of the Russia Federation (agreement contract № 14.575.21.0049 of 27.06.14).

References

[1] Krieger K, Geier A, Gong X, Maier H, Neu R, Rohde V 2003 *J. Nucl. Mat.* **313–316** 327

[2] Buzi L, De Temmerman G, Unterberg B, Reinhart M, Dittmar T, Matveev D, Linsmeier Ch, Breuer U, Kreter A, Van Oost G 2015 *J. Nucl. Mat.* **463** 320

[3] Krasheninnikov S I, Smirnov R D, Rudakov D L 2011 *Plasma Phys. Control. Fusion* **53** (8) 083001

[4] Begrambekov L, Buzhinskiy O 2006 *Problems of Atomic Science and Technology, Ser. Thermonuclear Fusion* **4** 14

[5] Azizov E A, Ayrapetov A A, Begrambekov L B et al 2014 *Problems of Atomic Science and Technology, Ser. Thermonuclear Fusion* **37** (4) 30

Journal of Physics: Conference Series **748** (2016) 012004 doi:10.1088/1742-6596/748/1/012004

The development and the tests of the electrostatic probe for dust particle collection in thermonuclear reactors

L B Begrambekov, A N Voityuk, A M Zakharov

National Research Nuclear University MEPhI (Moscow Engineering Physics Institute), 31 Kashirskoe highway, Moscow 115409, Russia

Corresponding author e-mail: lbb@plasma.mephi.ru

Abstract. Formation of dust particles in thermonuclear reactors can greatly affect the plasma parameters and lead to accumulation of tritium. The rates of formation and deposition of dust need to be measured, and the parameters of formation of dust particles and clusters need to be studied. A model of a device for collection of fine conductive particles capable of removing them from the reactor chamber for future research is proposed in this paper. The dust collector's operation is based on a principle of applied electrostatic field. The model was tested in different operating conditions: in vacuum, at the atmospheric pressure in the atmosphere of air and dry nitrogen. The experiments were conducted with a stationary system and with the dust collector in motion relative to the dusty surface. It is shown that, during the probe moving relative to the surface, it can remove up to 95% of fine tungsten particles with sizes ranging from 1 to 10 μm.

1. Introduction

Formation of dust particles and clusters has been observed in almost every modern thermonuclear reactor. Accumulation of dust in the next-generation thermonuclear reaction chambers can have a strong effect on plasma parameters and lead to accumulation of unacceptable amounts of tritium. For this reason, a device able to collect dust formed during the work of reactor and extract it from the chamber for investigation is necessary.

A model of such device (electrostatic dust collector) is proposed in this paper, a working model is produced and tested. During the trials the model of the dust collector was used for the removal of fine metallic particles composed of tungsten in different conditions: in vacuum, in air and in the atmosphere of dry nitrogen gas. The experiments were conducted for the dust collector in stationary condition and in motion relative to the dusty surface. It is shown that during the movement of the dust collector relative to the surface covered with fine particles with the size range of 1-10 μm, accumulation of up to 95% of particles is achievable.

2. On the possibility of the removal of metallic dust from the surface via applied electric field

The behaviour of fine tungsten dust particles in electrostatic field was investigated in [1]. For the experiments, tungsten powder with W content ≥ 99.5% and particle size between 1 and 10 μm was used. The size of the particles of tungsten powder was comparable to the size of dust particles formed in the chambers of modern thermonuclear reactors [2]. Additionally, investigation of powder under electron microscope has lead to a discovery of most powder particles forming clusters of different shapes and sizes of up to approximately 50 μm (figure 1).

Figure 1. SEM photograph of the tungsten powder particles.

The experiments on the effects of the electrostatic field on tungsten powder were conducted in vacuum at room temperature. The layer of tungsten powder was located on a horizontal plate. A second plate was located parallel to the first plate and above it, then an electric field was generated between the two plates. When the field was applied, the powder particles were also biased towards the upper plate and were attracted to it. It was shown that with the intensity of the electric field equaling to 600 V/mm, an emission of powder particles between the plates going from the lower electrode to the upper one occurred independent of the direction of the electric field. Additionally, the electric field intensity value was enough to remove both powder particles of all sizes and particle clusters from the lower plate. The rate of emission was increasing with the increase of the electric intensity. This phenomenon was proposed as a technique to clean dust-covered surfaces.

3. Separate deposition of zirconium and yttrium at floating potential on the substrate

A model of a dust collector was prepared. The main element of a model is a reservoir made of a thin sheet of stainless steel, with two sides being horizontal rectangular plates positioned one above the other. The plates are bounded by vertical walls from three sides, with the fourth wall made as a grid and tilted relative to the surface. The width of the horizontal sides of a dust collector was 25 mm, and the height of a probe was 10 mm. A potential difference enough for fine particles to overcome gravitational pull and adhesion was applied between the dust-covered surface and the collector. The fine particles then entered the collector's reservoir by going through the grid in the tilted wall. During the movement of a dust collector parallel to the surface with the gridded wall being in the front during the movement, a particle removal from the surface was conducted (figure 2).

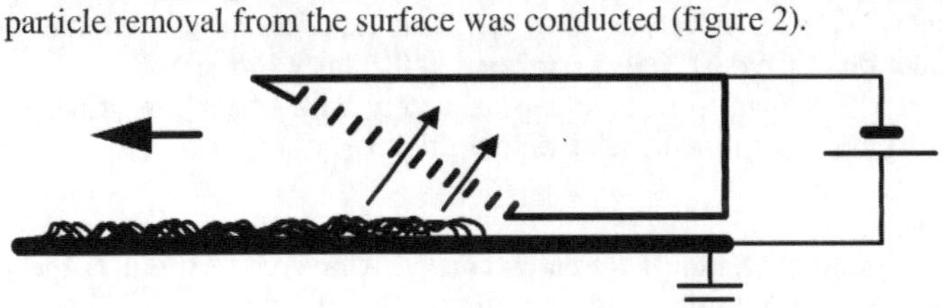

Figure 2. The scheme of the operating electrostatic dust collector.

4. Joint deposition of zirconium and yttrium at floating potential on the substrate

The dust collector was tested both in vacuum with pressure of approximately $1 \cdot 10^{-4}$ Torr and at the atmospheric pressure. The gases used in the latter trials were air and dry nitrogen.

Journal of Physics: Conference Series **748** (2016) 012004 doi:10.1088/1742-6596/748/1/012004

A series of tests was run in vacuum with a stationary dust collector. The dust particles formed a pile with the diameter of up to 10 mm and a height of approximately 1 mm, located on a grounded metal surface under the tilted wall of the dust collector. The interval between the bottom of the collector and the dust-covered surface was 3 mm. When the bias voltage above approximately 2000 V was applied to the dust collector, most of the fine particles disappeared from the surface and entered the reservoir in about 6-8 minutes. The rate of dust collection and the quantity of particles removed increased with the increase of bias voltage of the collector. The sign of the bias on the collector did not affect the rate of transfer of dust from the pile into the reservoir. At the potential difference of 5000 V, the highest value used in the experiment, the maximum amount of particles collected in the reservoir was 86% for the dust collector with the gridded wall at 30° relative to the surface. About 3% of the powder was left in place, and 5% stuck to the outer surface of the dust collector. The remaining 6% of the powder was scattered sideways in the gap between the surface and the collector.

The results of the tests conducted in the atmosphere of dry nitrogen were comparable with the results obtained in vacuum. For air, increased adhesion of powder to the outer surface of the dust collector as well as increased adhesion to the grounded surface was observed. Accumulation of particles did not exceed 60-65%

In the conditions of the experiment, motion of the collector relative to the surface was achieved by moving the grounded plate with the powder while keeping the collector stationary, as this was the most convenient option. The dust collector with bias voltage of 5000 V was fixed in place, and the powder was located on a grounded plate and moving towards the collector. The powder was poured on the plate in a line up to 50 mm in length. The width of the line varied for the purpose of investigating the properties of the removal of fine particles located at different distances from the axis of the collector. An optimization of the shape of the collector's gathering part was made based on the results of the experiments, which allowed for the reduction of the amount of fine particles scattered and stuck to the collector's exterior and practically removed the difference between the removal of fine particles from central and peripheral parts of the collector. Additionally, it was found that during the movement of the collector relative to the powder the amount of dust left on the grounded plate is greatly decreased due to all the fine particles are nearing the dust collector in the area of the highest electric field intensity. As a result, the experiments had shown that permanent losses of dust due to scattering could be reduced to about 5% of the total mass of particles, and, additionally, the direction of further improvement of the dust collector was found.

5. Conclusion

A model of an electrostatic dust collector for removing fine metal particles in thermonuclear reactor chambers was developed and produced.

The effectiveness of the dust collector model for removing fine metal particles composed of tungsten from flat horizontal surface was tested in various conditions: in vacuum, at the atmospheric pressure with atmospheres of air and dry nitrogen. The experiments were conducted in a stationary system and with motion of the dust collector relative to the dusty surface.

The dust collector is shown to remove approximately 95% of tungsten fine particles with sizes ranging from 1 to 10 μm while moving relative to the dusty surface.

Acknowledgments

This work was supported by National Research Nuclear University MEPhI in the framework of the Russian Academic Excellence Project (contract No. 02.a03.21.0005, 27.08.2013) and by the Ministry of Education and Science of the Russian Federation (agreement contract №14.575.21.0049 of 27.06.14).

References

[1] Begrambekov L et al. 2015 *Nucl. Instrum. Methods Phys. Res. B* **354** 282
[2] Krasheninnikov S I et al. 2011 *Plasma Phys. Controlled Fusion* **53** 083001

Journal of Physics: Conference Series **748** (2016) 012005 doi:10.1088/1742-6596/748/1/012005

Determination of atomic hydrogen in hydrocarbons by means of the reflected electron energy loss spectroscopy and the X-ray photoelectron spectroscopy

V P Afanas'ev[1], A S Gryazev[1], D S Efremenko[2], P S Kaplya[1], O Y Ridzel[1]

[1]National Research University "Moscow Power Engineering Institute", Department of General Physics and Nuclear Fusion, 14 Krasnokazarmennaya st., Moscow 111250, Russia

[2]Deutsches Zentrum für Luft- und Raumfahrt (DLR), Institut für Methodik der Fernerkundung (IMF), Oberpfaffenhofen 82234, Germany

Corresponding author e-mail: pavel@kaplya.com

Abstract. Elastic peaks electron spectroscopy (EPES) is a perspective tool for measuring the hydrogen atomic density in hydrocarbons. It is known that hydrogen elastic peaks overlap inelastic energy loss spectra. This fact complicates the quantitative interpretation of EPES spectra. In this paper, a novel technique based on the joint use of EPES and X-ray photoelectron spectroscopy (PES) is proposed. A key part of the method is the inelastic scattering background subtraction which is performed in two steps. At the first step, differential inelastic scattering cross-sections are retrieved from PES spectra, while at the second step, the retrieved cross-sections are used to remove the inelastic scattering signal from EPES spectra. Both REELS and PES spectra are described on the base of the invariant imbedding method forming a consistent framework for the surface state analysis. A good agreement is obtained between calculated spectra and experimental data.

1. Introduction

Hydrogen influences mechanical properties of construction materials. The studies of hydrogen storage in metals and plasma-material interactions in a fusion reactor require an efficient method for quantitative analysis of hydrogen isotopes. According to [1], methods based on the electron spectroscopy usually have problems with detecting hydrogen and helium. Nevertheless, recently several techniques have been proposed for measuring the bound hydrogen by means of electron spectroscopy [2–11]. A perspective tool for hydrogen detection is the elastic peak electron spectroscopy (EPES) [10, 12]. In the EPES, information on the sample composition is acquired from elastic energy losses (typically, in the range 0-20 eV) which are due to the elastic electron reflection. The amount of electron energy loss ΔE in a single elastic scattering event can be computed using the conservation laws for energy and momentum. The final expression reads:

$$\Delta E \cong \frac{2m}{M} E_0 (1-\cos\psi) \quad (1)$$

Here m is the electron mass, M is the scattering atom mass, E_0 is the initial electron energy and ψ is the scattering angle. Elastically scattered electrons which have not experienced inelastic scattering

Journal of Physics: Conference Series **748** (2016) 012005 doi:10.1088/1742-6596/748/1/012005

processes form a so-called "elastic peak" in the spectrum. The peak shape can be approximated by the Gaussian distribution [13] with full width at half maximum (FWHM) given by

$$2\sigma = 2\sqrt{\sigma_A^2 + \sigma_B^2 + \sigma_D^2} \tag{2}$$

where σ_A is the broadening due to the energy analyzer slit function, σ_B is the broadening of the electron beam, σ_D is the Doppler broadening. The Rayleigh criterion for two peaks with maximums at E_1 and E_2 to be distinguished reads as $|E_1 - E_2| \geq \sigma$. From (1) it follows that the hydrogen atoms with the lowest atomic mass possible have the largest elastic peak shift. Therefore, elastic peaks corresponding to hydrogen isotopes can be easily resolved on the state-of-the-art setups. However, the hydrogen elastic peak overlaps with the inelastic scattering losses. This complicates conducting analysis of EPES data.

In this paper, we propose a method for the inelastic scattering background subtraction from EPES spectra. This method is based on the joint analysis of the photoelectron spectra (PES) and REELS spectra. Essentially, any mathematically justified technique of the background subtraction has to utilize information on the differential inverse inelastic mean free path (DIIMFP):

$$\omega_{in}(\Delta) = \frac{d\sigma_{in}}{d\Delta} \tag{3}$$

where $\sigma_{in} = (nl_{in})^{-1}$ is the inelastic electron scattering cross-section and l_{in} is the inelastic mean free path (IMFP). Therefore, the background subtraction is performed in two steps. At the first step, differential inelastic scattering cross-sections are retrieved from PES spectra, while at the second step, the retrieved cross-sections are used to remove the inelastic scattering signal from EPES spectra.

2. Theoretical

2.1. General considerations

We consider the hydrocarbon surface samples that has been prepared in the Max Planck Institute of Plasma Physics (Max-Planck-Institut für Plasmaphysik, IPP) on the plasma-chemical sputtering setup [10]. The samples are investigated using the experimental setup NanoFab 25 at the National Research University «MPEI». The initial analysis is performed using the XPS method. The corresponding PES spectrum is shown in figure 1. Apparently, the sample is quite clean since almost only those peaks that correspond to O and C are seen in the spectrum. Note, that the spectrum contains a background, which is due to inelastic scattering events (plasmon excitations and ionization of the inner electron shells).

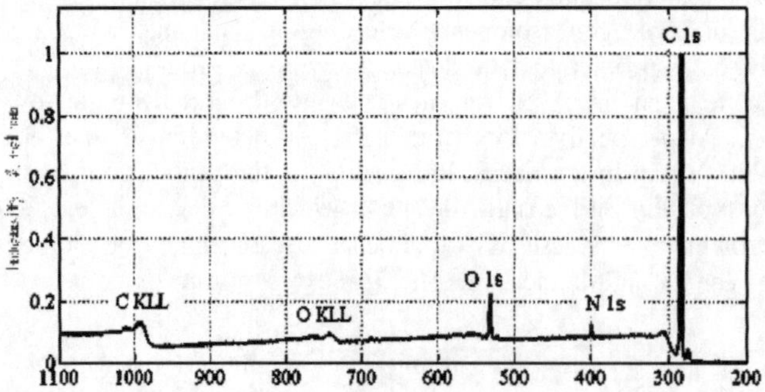

Figure 1. The PES spectrum for the hydrocarbon: excitation is done using the nonmonochromatic Mg K_α X-ray source.

Given the initial electron energy of 3 keV, the value of energy losses after one elastic scattering collision by a hydrogen atom according to Eq. (1) equals to 4.88 eV. Essentially, $\omega_{in}(\Delta)$ is nonmonotonic function with the first maximum in the region 10-20 eV. In the region 5-10 eV, $\omega_{in}(\Delta)$ increases with Δ and provide the signal comparable to the elastic peaks. The shape of $\omega_{in}(\Delta)$ which is referred to as a normalized differential inverse inelastic mean free path (NDIIMFP) is given by

$$x_{in}(\Delta) = \frac{\omega_{in}(\Delta)}{\sigma_{in}}, \quad \int_0^{E_0} x_{in}(\Delta)d\Delta = 1 \tag{4}$$

Below, we describe an algorithm for NDIIMFP retrieval from PES spectra.

2.2. *Accounting for multiple scattering*

Mathematically, the NDIIMFP retrieval problem is a deconvolution problem. To perform the deconvolution, we have to map the single scattering signal to the multiple scattering signal. To account for the impact of multiple scattering collisions, we adopt the formalism of Ambartsumian-Chandrasekhar equations. Let us introduce functions Q_k, R_k which are the photo-electron flux density function and the reflection function of particles that have experienced exactly k inelastic scattering events, respectively. Then, in the wide range of energy losses the functions Q and R are expanded in series of Q_k- and R_k-functions, respectively:

$$Q(\Delta,\mu_0,\mu,\varphi) = Q_0(\mu_0,\mu,\varphi)\delta(\Delta) + \sum_{k=1}^{\infty}\left[Q_k(\mu_0,\mu,\varphi)x_{in}^k(\Delta)\right] \tag{5}$$

$$R(\Delta,\mu_0,\mu,\varphi) = R_0(\mu_0,\mu,\varphi)\delta(\Delta) + \sum_{k=1}^{\infty}\left[R_k(\mu_0,\mu,\varphi)x_{in}^k(\Delta)\right] \tag{6}$$

$$x_{in}^1(\Delta) = x_{in}(\Delta), \quad x_{in}^{k+1}(\Delta) = \int_0^{\Delta} x_{in}^k(\Delta-\varepsilon)x_{in}(\varepsilon)d\varepsilon \tag{7}$$

Here Q_k and R_k correspond to k-fold inelastically scattered particles. If $x_{in}(\Delta)$ stands for the energy spectrum, which is observed after a single inelastic scattering event, then the k-fold convolution $x_{in}^k(\Delta)$ identifies the energy loss spectrum after k successive inelastic scattering events. Essentially, $Q_k(\mu_0,\mu,\varphi)$ can be found only together with $R_k(\mu_0,\mu,\varphi)$. The equations for $Q_k(\mu_0,\mu,\varphi)$ and $R_k(\mu_0,\mu,\varphi)$ obtained using the invariant imbedding method are given in [14, 15]. The derivation consists in adding a layer of thickness $d\tau$ above the τ-thickness layer and finding corresponding changes in Q_k and R_k. The value of $d\tau$ is assumed to be small enough to neglect multiple scattering processes in this layer. The equation for $Q_k(\mu_0,\mu,\varphi)$ for a semi-infinite layer is given by [14, 16]

$$\begin{aligned}&\frac{1}{\mu}Q_k(\mu_0,\mu,\varphi) - (1-\delta_{k0})\frac{1-\lambda}{\mu}Q_{k-1}(\mu_0,\mu,\varphi) = \delta_{k0}\lambda_\gamma f(\mu_0,\mu,\varphi)\\&+\lambda_\gamma f\otimes R_k + \lambda Q_k\otimes x_{el} + \lambda Q_k\otimes x_{el}\otimes R_0 + \lambda\sum_{j=0}^{k-1}Q_j\otimes x_{el}\otimes R_{k-j}\end{aligned} \tag{8}$$

Here $\lambda = \dfrac{\sigma_{el}}{\sigma_{el}+\sigma_{in}}$, $\lambda_\gamma = \dfrac{\sigma_\gamma}{\sigma_{el}+\sigma_{in}}$, σ_{el} is the total elastic cross-section [17] and σ_γ is the total photo-ionization cross-section [18]. The values of l_{in} are taken from [19].

The equation for $R_k(\mu_0,\mu,\varphi)$ for a semi-infinite layer was considered in [15]. It reads:

$$\left(\frac{1}{\mu}+\frac{1}{\mu_0}\right)R_k = \lambda x_{el}\delta_{k0} + \lambda x_{el}\otimes R_k + \lambda R_k \otimes x_{el} + \lambda R_0 \otimes x_{el} \otimes R_k$$
$$+\lambda R_k \otimes x_{el} \otimes R_0 + \lambda \sum_{j=1}^{k-1} R_j \otimes x_{el} \otimes R_{k-j} + (1-\lambda)\left(\frac{1}{\mu}+\frac{1}{\mu_0}\right)R_{k-1} \quad (9)$$

In the context of Eqs. (8) and (9), the samples are assumed to be homogeneous, which implies that the difference between surface and bulk scattering properties is neglected. Further, the NDIIMFP is approximated as follows:

$$x_{in}(\Delta) = \sum_{i=1}^{N_{pl}} \lambda_{pl\,i} x_{pl\,i}(\Delta) + \sum_{j=1}^{N_{ion}} \lambda_{ion\,j} x_{ion\,j}(\Delta)$$
$$x_{pl\,i}(\Delta) = \frac{A_{pl\,i}\Delta^{\beta}}{\left(\Delta^2 - \varepsilon_{pl\,i}^2\right)^2 + \Delta^{\alpha} b_i^{4-\alpha}}, \quad x_{ion\,j}(\Delta) = \frac{A_{ion\,j}}{\Delta^2}\eta\left(\Delta - J_{ion\,j}\right) \quad (10)$$

where $J_{ion\,j}$ is the ionization threshold and $\eta\left(\Delta - J_{ion\,j}\right)$ is the Heaviside step function for the ionization energy. The coefficients $A_{pl\,i}$ and $A_{ion\,j}$ are introduced here to fulfill the following normalization conditions:

$$\int_0^{E_0} x_{pl\,i}(\Delta)\,d\Delta = 1, \quad \int_0^{E_0} x_{ion\,j}(\Delta)\,d\Delta = 1$$
$$\sum_{i=1}^{N_{pl}} \lambda_{pl\,i} + \sum_{j=1}^{N_{ion}} \lambda_{ion\,j} = 1$$

Essentially, Eqs. (8) and (9) are Sylvester linear matrix equations, while Eq. (9) setting k=0 is the Ricatti matrix equation. These equations are discretized in the angular domain and solved numerically in the discrete ordinate space using standard mathematical packages. Note, that the computational time for solving Eqs. (8) and (9) does not exceed a second. Therefore, this technique can be efficiently used in the fitting procedure.

3. Results and discussion

To compare calculations to experimental data, the result from Eq. (5) should be convoluted with the slit function $G_Q(\varepsilon)$ as follows:

$$Q_{fit}(\Delta,\mu_0,\mu,\varphi) = \int_0^{\Delta} Q(\Delta-\varepsilon,\mu_0,\mu,\varphi)G_Q(\varepsilon)\,d\varepsilon \quad (11)$$

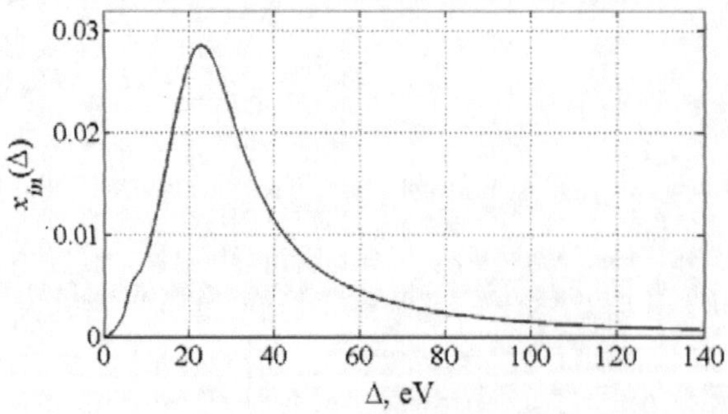

Figure 2. The NDIIMFP retrieved from the PES spectrum shown in figure 1.

The coefficients α, β, b_i, $\varepsilon_{pl\ i}$, $J_{ion\ j}$ in Eq. (10) are found in the fitting procedure by minimizing the following objective function:

$$\gamma = \int_0^{\Delta_{max}} \left| Q_{exp}(\Delta,\mu_0,\mu,\varphi) - Q_{fit}(\Delta,\mu_0,\mu,\varphi) \right| d\Delta \tag{12}$$

where (0, Δ_{max}) is the considered range of energy losses.

The comparison of the calculated spectrum using the retrieved NDIIMFP with the experimental spectrum is shown in figure 3, while the reconstructed NDIIMFP is presented in figure 2. The carbon oxide peak and the Mg $K_{\alpha 3}$, Mg $K_{\alpha 4}$ X-ray source satellite peaks are taken into account when computing Q_{fit}. The oxide peak in figure 3 is shifted from the main elastic peak by 3.6 eV. The relative intensities of the primary $\alpha_{1,2}$ and satellite α_3, α_4 X-ray peaks are given in Table 1. The corresponding data is taken from [1]. Given the reconstructed cross-section, the REELS spectrum $R(\Delta,\mu_0,\mu,\varphi)$ is computed and compared to the experimental data. The initial energy is 3 keV. Note, that just as for $Q(\Delta,\mu_0,\mu,\varphi)$, $R(\Delta,\mu_0,\mu,\varphi)$ should be convoluted with the slit function $G_R(\varepsilon)$ as follows:

$$R_{fit}(\Delta,\mu_0,\mu,\varphi) = \int_0^{\Delta} R(\Delta-\varepsilon,\mu_0,\mu,\varphi) G_R(\varepsilon) d\varepsilon \tag{13}$$

The comparison is illustrated in figure 4.

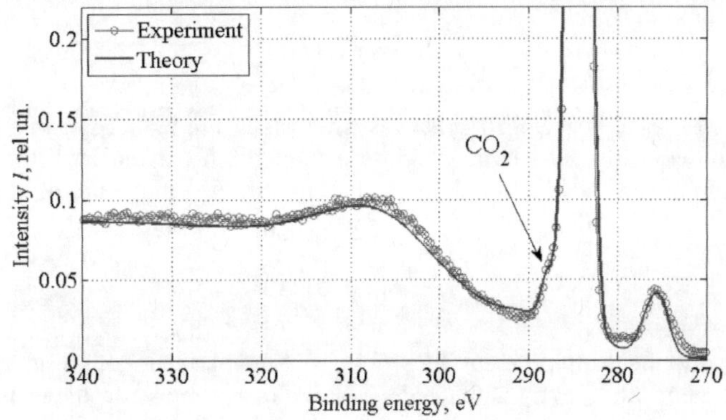

Figure 3. The PES spectrum of the sample: excitation is done by using the nonmonochromatic Mg K_α X -ray source.

Table 1. X-ray satellite energies and intensities for Mg source.

	$\alpha_{1,2}$	α_3	α_4
Rel. intensity (%)	100	8.0	4.1
Energy displacement (eV)	0	8.4	10.2

The misfit between the experimental data and $R_{fit}(\Delta,\mu_0,\mu,\varphi)$ in the energy range of 2990÷2998 eV is due to the electron energy losses which occur during the elastic scattering collisions with hydrogen atoms. This effect in the context of EPES spectroscopy was quantitatively studied in [11, 13 20]. The FWHM of the EPES hydrogen peak agrees well with Eq. (2).

Journal of Physics: Conference Series **748** (2016) 012005 doi:10.1088/1742-6596/748/1/012005

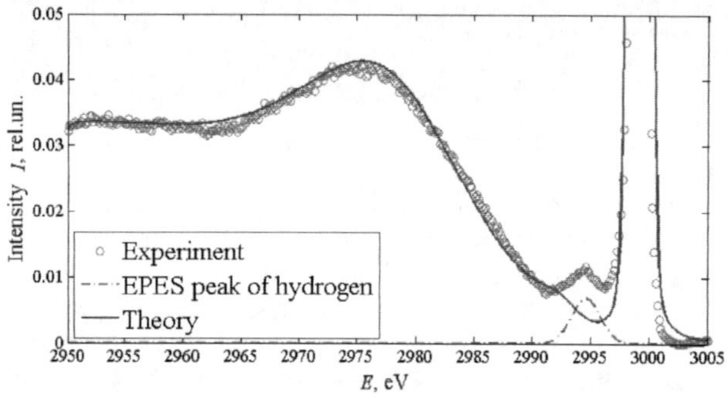

Figure 4. Comparison of the experimental data (circles) [10] and the calculated REELS spectrum using reconstructed cross-section (solid line). The dash-dot line corresponds to the retrieved EPES signal after subtracting the inelastic scattering background.

To perform a quantitative analysis, we assume that the investigated sample is homogeneous within a layer of thickness of l_{in}=6.5 nm, which corresponds to the EPES information depth [27]. With this assumption, we have a bijection between the elastic peak intensity and the quantity of the related element [21]. The fitting procedure gives the following ratio:

$$\frac{n_H}{n_C} = 1.3 \pm 0.1$$

Note, that this retrieved ratio is averaged in some sense over 10 nm layer from which the EPES signal originates from. Consequently, by varying the information depth our retrieval results can be refined and the depth resolution can be enhanced. In particular, in [10] the sample has been investigated using the nuclear reaction analysis (NRA) and the following ratio has been measured:

$$\frac{n_H}{n_C} = 0.53 \pm 0.04$$

This ratio is valid up to the depth of about 1 μm [22, 23]. Summarizing both results, we conclude that the amount of hydrogen is decreasing with the depth. For a more precise depth profiling, several EPES spectra measured at several initial energies E_0 (providing different information depths) can be incorporated in the presented algorithm.

4. Conclusion

The quantitative composition analysis of solids can be performed by using the elastic peak electron spectroscopy [3, 21]. The intensity of the elastic peaks depends on the concentration depth profile of a certain element [24]. In this paper, we proposed a method for the inelastic scattering background removal. This method is based on the joint interpretation of the REELS and PES spectra using the same NDIIMFPs retrieved from PES spectra. The background subtraction is performed in two steps. At the first step, differential inelastic scattering cross-sections are retrieved from PES spectra, while at the second step, the retrieved cross-sections are used to remove the inelastic scattering signal from EPES spectra.

Good agreement is obtained between the experimental data and numerical computations using Eqs. (8) and (9). From this we can conclude that the physical model that has been used for derivation of these equations is valid. In particular, there is a controversy regarding the photoelectron energy losses due to so-called intrinsic plasmon excitations. In this paper, all computations of electron energy loss spectra have been performed without accounting for intrinsic plasmon excitations, while all physically relevant processes such as ionization, plasmon excitation and elastic collisions are considered in the same manner both for REELS and PES. The absence of intrinsic excitations was stated also in [14, 25, 26] in the context of the similar investigation of Al, Si, Mg and Nb samples. The presented NDIIMFP reconstruction technique accompanied with numerical tools for simulating REELS spectra leads to a consistent elastic peak analysis. Unlike several techniques described in [9], the proposed methods can be implemented using just one experimental setup with both REELS and PES capabilities.

References

[1] Hofmann S 2013 *Auger- and X-Ray Photoelectron Spectroscopy in Materials Science (Springer Series in Surface Sciences* vol 49) (Berlin, Heidelberg: Springer Berlin Heidelberg)
[2] Vos M 2001 *Phys. Rev.* A **65** 012703
[3] Vos M 2002 *Ultramicroscopy* **92** 143
[4] Vos M, Chatzidimitriou-Dreismann C, Abdul-Redah T and Mayers J 2005 *Nucl. Instrum. Methods Phys. Res., Sect. B* **227** 233
[5] Sulyok A, Gergely G, Menyhard M, Toth J, Varga D, Kover L, Berenyi Z, Lesiak B and Kosinski A 2001 *Vacuum* **63** 371
[6] Gergely G 2002 *Prog. Surf. Sci.* **71** 31
[7] Orosz G, Gergely G, Menyhard M, Tóth J, Varga D, Lesiak B and Jablonski A 2004 *Surf. Sci.* **566-568** 544
[8] Yubero F, Rico V J, Espinos J P, Cotrino J and Gonzalez-Elipe A R 2005 *Appl. Phys. Lett.* **87** 084101
[9] Yubero F and Tokesi K 2009 *Appl. Phys. Lett.* **95** 084101
[10] Kostanovskiy I, Afanas'ev V, Naujoks D and Mayer M 2015 *J. Electron. Spectrosc. Relat. Phenom.* **202** 22
[11] Afanas'ev V P, Afanas'ev M V, Lisov A A and Lubenchenko A V 2009 *Tech.Phys.* **54** 1667
[12] Vos M and Went M 2007 *J. Electron. Spectrosc. Relat. Phenom.* **155** 35
[13] Afanas'ev V, Afanas'ev M, Lubenchenko A, Batrakov A, Efremenko D and Vos M 2010 *J. Electron. Spectrosc. Relat. Phenom.* **177** 35
[14] Afanas'ev V, Golovina O, Gryazev A, Efremenko D and Kaplya P 2015 *J. Vac. Sci. Technol. B* **33** 03D101
[15] Afanas'ev V P and Kaplya P S 2015 *J.Surf.Invest.:X-ray, Synchrotron Neutron Tech.* **9** 715
[16] Afanas'ev V, Kaplya P, Lubenchenko A and Lubenchenko O 2014 *Vacuum* **105** 96
[17] Jablonski A, Salvat F and Powell C 2010 *NIST Electron Elastic-Scattering Cross-Section Database* – Version 3.2 (Gaithersburg, MD: National Institute of Standards and Technology)
[18] Trzhaskovskaya M, Nefedov V and Yarzhemsky V 2001 *At. Data Nucl. Data Tables* **77** 97
[19] Tanuma S, Powell C and Penn D 2011 *Surf. Interface Anal.* **43** 689
[20] Afanas'ev V P, Afanas'ev M V, Batrakov A A, Bohmeyer W, Naujoks D, Lubenchenko A V and Markin A 2011 *J. Surf. Invest.: X-ray, Synchrotron Neutron Tech.* **5** 70
[21] Afanas'ev V, Efremenko D and Lubenchenko A 2013 On the application of the invariant embedding method and RTE codes for surface state analysis *Light Scattering Reviews 8*, ed A A Kokhanovsky (Berlin, Heidelberg: Springer Berlin Heidelberg) chapter 8 pp 363
[22] Ziegler J F and Biersack J P 1985 The stopping and range of ions in matter *Treatise on Heavy-Ion Science*, ed D Bromley (Boston, MA: Springer US) 93
[23] Feldman L and Picraux S 1977 Selected low energy nuclear reaction data *Ion Beam Handbook for Material Analysis*, ed J Mayer and E Rimini (New York: Academic Press) chapter 4
[24] Afanasyev V P, Efremenko D S, Lubenchenko A V, Vos M, Went M R 2010 *Bull. Russ. Acad. Sci.: Phys.* **74** 170
[25] Afanas'ev V P, Gryazev A S, Kaplya P S, Andreyeva Y O, Golovina O Y 2016 *J. Surf. Invest.: X-ray, Synchrotron Neutron Tech.* **10** 101
[26] Afanas'ev V P, Gryazev A S, Kaplya P S, Andreyeva Y O 2016 *J. Surf. Invest.: X-ray,*

Journal of Physics: Conference Series **748** (2016) 012005 doi:10.1088/1742-6596/748/1/012005

Synchrotron Neutron Tech. **10** 113
[27] Jablonski A, Powell C J 2004 *Surf. Sci.* **551** 106

Journal of Physics: Conference Series **748** (2016) 012006 doi:10.1088/1742-6596/748/1/012006

Oleophobic optical coating deposited by magnetron PVD

D Bernt[1], V Ponomarenko[1] and A Pisarev[2]

[1]NSG Pilkington Glass, 1 Stekolnaya st., Zhukovo village Chulkovsky s/o Ramensky District, Moscow region 140125, Russia
[2]National Research Nuclear University MEPhI (Moscow Engineering Physics Institute), 31 Kashirskoe highway, Moscow 115409, Russia

Corresponding author e-mail: Dmitriy.Bernt@spglass.ru

Abstract. Thin oxinitride films of Zn-Sn-O-N and Si-Al-O-N were deposited on glass by reactive magnetron sputtering at various nitrogen-to-oxygen ratios. Nitrogen added to oxygen led to decrease of the surface roughness and increase of oleophobic properties studied by the oil-drop test. The best oleophobity was obtained for Zn-Sn-O-N oxinitride at Zn:Sn=1:1 and N:O=1:2. Improved oleophobic properties were also demonstrated if the oxinitride film was deposited on top of the multilayer coating as the final step in the industrial cycle of production of energy efficient glass.

1. Introduction

Sputter deposited thin films are widely used in production of architectural flat glass [1]. These coatings consist of 3 to 40 layers of various materials mainly of metals and metal oxides. The thickness of individual layers is in the range of several to hundred nanometers [2]. The energy efficient glasses, for example, consist of a sequence of dielectric and silver layers, where thin (~10 nm) silver layers reflect infrared radiation back into the room, while dielectric layers protect the silver from chemical and mechanical degradation and are anti-reflecting [3]. All the layers are usually deposited by magnetron sputtering, which is a standard coating technology for large-area flat substrates [4-7].

A common problem of these energy-efficient coatings is their wear damage and surface fouling due to deposition of non- or hardly- removable oily and suety contaminations. This may happen due to contact with liquids for glass-cutters, transport rollers, dirty protective gloves, bare hands, etc.

To overcome this problem, considerable efforts have been undertaken to develop a technology of the oleophobic surface. Surface oleophobity states for the surface tension to repel oils and viscous liquids. It leads to a diverse array of applications including self-cleaning, stain-free cloth and drag reduction. It is known that oleophobic properties are improved if the surface is uniformly nanostructurized and the surface curvature of its elements is big. Such surface structure has large contact angles with viscous liquids (above 150 degrees) and small contact angle hysteresis. Materials free of pores and cavities, which could act as nanocapillars, demonstrate a better oleophobity [8]. The methods used for oleophobic processing are: electrospinning, liquid vapor deposition and spray atomization [8-10]. They however cannot be applied to thin layers.

The purpose of this study is to analyze the possibility of improving oleophobic properties of thin films in application to energy-efficient transparent conductive multilayer coatings on glass substrates.

Journal of Physics: Conference Series **748** (2016) 012006 doi:10.1088/1742-6596/748/1/012006

The process must be easily incorporated in the technological cycle of PVD magnetron coating and must not affect negatively the properties of the main functional coating.

The dielectric materials mainly used as the top layers of transparent conductive energy-efficient coating on glasses are two-component metal oxides where metals are from the group: Ti, Si, Zn, Sn, In, Zr, Al, Cr, Nb, Mo, Hf, Ta, and W. The oxides are good barriers against oxygen diffusion and withstand mechanical stresses [2] thus preventing chemical degradation, cracking, and exfoliation. The surface of the outer oxide layer needs protection against oily contaminations.

It was observed, that introduction of nitrogen as a second gas component along with oxygen during sputter deposition favors formation of a more dense and uniform nanostructurized surface structure with high surface curvature of its elements [11]. Therefore, oxinitrides seem to be promising from the point of view of improving oleophobity. Two metal compounds for oxinitriding were selected: Zn-Sn and Si-Al, as the respective oxide layers are often used in multilayer optical coatings on architectural glasses [12-16].

2. Experimental

Deposition of protective coatings was performed in an industrial line VonArdenne GC330H for coating of architectural glass with dimensions of 4×6 m. This line is used for multi-layer deposition of metals and metal oxides. Full-scale soda-lime glass sheets 4 mm thick were used for deposition. The sheets were cleaned according to a standard technological procedure before film deposition. This procedure was a sequence of cleaning in deionized water and drying using compressed dry air. Films of zinc-stannate oxinitride Zn-Sn-O-N, and aluminum doped silica oxinitride Si-Al-O-N were physically deposited by reactive magnetron sputtering directly on bare glass substrates without intermediate layers. Glass sheets were not heated during deposition.

After experiments with bare glass sheets, the optimal oleophobic coating was also deposited on top of the multilayer energy-efficient industrial glass product by adding an extra step of oxinitride layer deposition to the standard technological chain.

Targets for sputtering were made of respective bi-metal alloys. Two Zn-Sn targets with various alloy compositions were used: one with 50 wt.% Sn (±2 wt.%) and another with 10 wt.% Sn (±2 wt.%). The Si-Al had 10 wt.% Al (±2 wt.%). The minimal purity of all the metals was 99.7%. Dual rotating cylindrical magnetron system, driven by a TRUMPF Hüttinger TruPlasma power supply was used. The power supply operated in pulsed DC mode at the frequency of 37 kHz and the power of 50 to 52 kW.

The amount of gases introduced into the sputtering chambers was controlled by the MKS mass flow controllers. Reactive sputtering has been performed in the mixture of argon, oxygen, and nitrogen. Argon was used as the sputter component, while oxygen and nitrogen were reactive components. Nitrogen flow varied from 0 to 800 sccm. Oxygen flow was set to sputter the targets in the transition mode of material poisoning hysteresis and for all the used materials was at the average of 800 sccm. The argon flow was set to keep the total pressure of the mixed gases at the level of approximately $3.5\cdot 10^{-3}$ mbar.

Comparison of the oleophobic properties was made by the so-called "oil-drop test". A small drop (0.5±0.1 ml) of a given viscosity was applied on the surface, and its diameter was measured in optical stereomicroscope. The drop diameter on the surface is directly proportional to the dip contact angle on the liquid/solid interface and characterizes oleophobity. The larger the diameter of the drop spread over the surface, the worse the oleophobic properties of the surface with respect to the liquid of given viscosity. Three liquids of different viscosity were used for testing: immersion oil for optical microscopy with the viscosity V = 100-120 mPa·s (T = 20°C), synthetic vacuum oil with V=60-70 mPa·s (T = 20°C), and liquid for glass-cutters with V = 20-40 mPa·s (T=20°C). In some cases the drop was round in shape and small in diameter, the contact angle was large, and this is the case of good oleophobity. In other cases the drop spread over the surface, its boundary was noncircular, the contact angle was small, and this is the case of bad oleophobity.

3. Results and Discussion

Results of the drop tests are shown in figure 1 for different test liquids. The diameter of the drop depends on viscosity of the test liquid and on the amount of nitrogen added to oxygen in the plasma-forming gas. In all cases nitrogen improves oleophobic properties of the surface.

The diameter of the drop as a function of the nitrogen flow demonstrates prominent decrease at the nitrogen flow rate above 200 sccm and has a minimum at around 350 sccm for all samples and all test liquids, which states for approximately a half of the oxygen flow (N:O=1:2). The rise of the diameter after the minima is rather slow.

The quantitative effect of nitrogen on oleophobity depends on the surface and test liquid used. Liquids with low viscosity are less sensitive to oleophobic treatment, while the effect is rather high for liquids with high viscosity. For example the radius decreases about 3 times in the case of oil with $V = 60\text{-}70$ mPa·s on Zn-Sn-O-N (at Zn:Sn=1:1). The best oleophobic result was demonstrated by oxinitride Zn-Sn-O-N with equal weight ratio of components in the Zn-Sn sputter target (Zn:Sn=1:1).

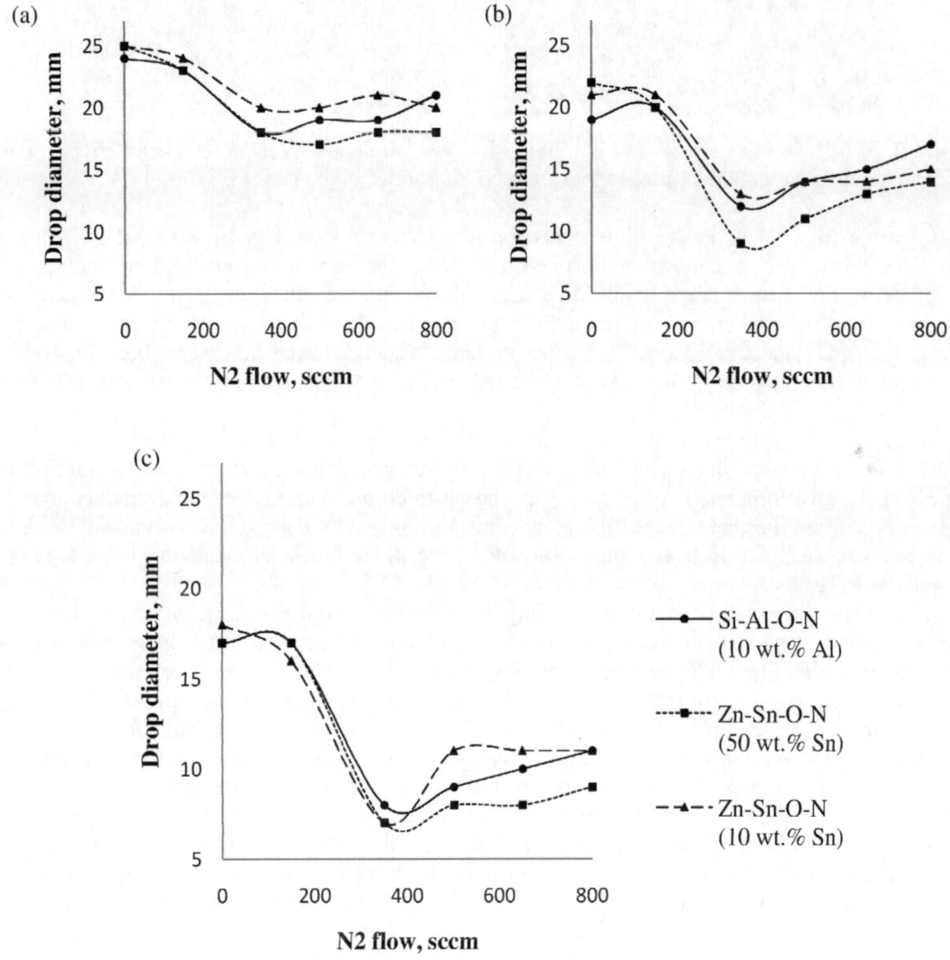

Figure 1. Results of the "oil-drop" tests of three protection coatings and three liquids of different viscosity: $V = 20\text{-}40$ mPa·s (a) ; $V = 60\text{-}70$ mPa·s (b); $V = 100\text{-}120$ mPa·s (c). Solid line Si-Al-O-N (10 wt.% Si), broken line Zn-Sn-O-N (10 wt.% Sn), dotted line Zn-Sn-O-N (50 wt.% Sn).

Examples of SEM images of surfaces of Zn-Sn-O and Zn-Sn-O-N (50 wt.% Sn both) is given in figure 2. Oxinitride films in comparison with pure oxides have a smoother surface. There are no cavities on the surface of oxinitride films, which can work as capillars and accumulate viscous liquids making it difficult to remove them.

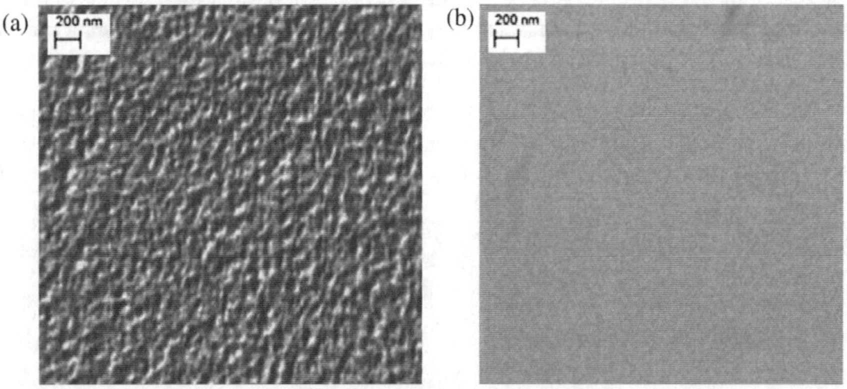

Figure 2. SEM images of surfaces of Zn-Sn-O (50% wt Sn) deposited by reacive sputtering in pure oxygen(a) and Zn-Sn-O-N deposited if 500 sccm N2 was added to oxygen (b).

Dimensions of crystals in the film depend on many factors. One may suggest that oxygen and nitrogen compete during deposition of the film, giving increase in variety and the number of precipitates of oxide, nitride, and oxinitride phases. Oleophobic properties are linked with dimensions of nanocrystals. The best oleophobic properties were obtained if the flow rates of nitrogen and that of oxygen were 1:2 respectively, and if the concentrations of metals in the film were about 50:50. One can speculate that maximal concentration of phase precipitates and minimal crystal dimensions are expected in these conditions.

The oleophobic layer with optimal properties Zn-Sn-O-N (Zn:Sn=1:1 and N:O=1:2) was then deposited as the top layer on the industrial multilayer energy-efficient glass sheet as a final stage of the single technological process. Its thickness was chosen to ensure that all the initial product spectral properties remained the same. Standard quality control tests used on the production line did not show any issues with chemomechanical sustainability after deposition of the additional oleophobic top layer to the standard layer stack.

The standard glass with a multilayer coating and the test glass with additional oleophobic layer were tested with respect to oil fouling. Glass cutting oil was spilled over coated glass, and in some areas the oil was smeared into the surface by dirty gloves. The surfaces were dried, and then brush-washed using the standard procedure in the industrial glass washing machine. After washing there were no traces of oil on the oleophobic surface, while the surface without oleophobic layer retained a lot of oil. The procedure of washing of the standard glass was repeated several times with increase of pressure on the washing brushes. It was found that sticking of the oil to the standard glass surface was so high that deposited layers were destroyed and partially removed before oil was washed out. Figure 3 shows two surfaces after washing: the oil-free surface with oleophobic layer after the very first washing and the oil-fouled surface without oleophobic layer after many cycles of washing. One can see obvious damage of the standard layer-stack, which happened before the oil had been completely removed.

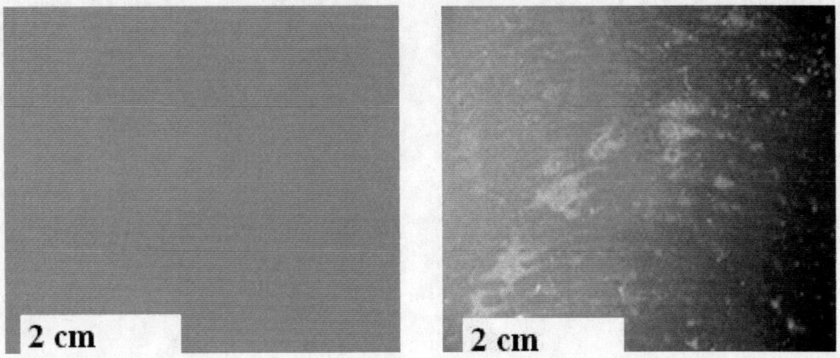

Figure 3. Optical microscopy of surfaces after brush-washing of the sample with oleophobic layer (left) and without it (right).

4. Conclusion

It was demonstrated that layers of oxynitrides Zn-Sn-O-N and Si-Al-O-N, which were deposited either on the glass surface directly or on the multilayer metal-dielectric stack on glass, both have prominent oleophobic properties. The oxinitrides have much smoother surface, which is possibly connected with nanocrystalline nature of the coating. The best result for Zn-Sn-O-N was obtained at Zn:Sn=1:1 and N:O=1:2, possibly because these conditions are preferential for nanocrystalline grain growth due to maximazing variety and number of phase precipitates.

Acknowledgments

The authors acknowledge Dr J.A. Ridealgh, Pilkington Technology Management Ltd., for support and discussion, as well as A. Atyubrin and A. Tikhonov, NSG Pilkington Glass, for oil fouling tests. The work was supported in part by Contract № 14.Y26.31.0008 from the Ministry of Education and Science of Russia and by National Research Nuclear University MEPhI in the framework of the Russian Academic Excellence Project (contract No. 02.a03.21.0005, 27.08.2013).

References

[1] E. Ando 2001 *Thin Solid Films* **392** 289
[2] Hackery K V, Meshvitz P A and Dannenberg R D 2003 *Patent RU* 2498954
[3] Kulczyk-Malecka J, Kelly P J, West G, Clarke G and Ridealgh J 2011 *Thin Solid Films* **520** 1368
[4] Weis H, Grosse P, Herlitze L, Friedrich I and Wuttig M 1999 *Thin Solid Films* **351** 184
[5] Finley J J 1999 *Thin Solid Films* **351** 264
[6] Ando E, Suzuki S, Aomine N, Miyazaki M and Tada M 2000 *Vacuum* **59** 792
[7] Hammarberg E 2003 *Thin Solid Films* **442** 222
[8] Steele A, Bayer I and Loth E 2009 *NanoLetters* **9** (1) 501
[9] Tuteja A, Choi W, Ma M, Mabry J M, Mazzella S A, Rutledge G C, McKinley G H, and Cohen R E 2007 *Science* **318** 1618
[10] Weber D and Matsuyuki N 2013 *US Patent Aplication* 61704275
[11] Qian J, Farokhzadeh K and Edrisy A 2014 *Surface & Coatings Technology* **258** 134
[12] Nikolić M V, Ivetić T, Paraskevopoulos K M, Zorba T T, Blagojević V, Mančić L and Nikolić P M 2008 *Thin Solid Films* **516** 6293
[13] Afanasyev-Charkin I V, Jacobsohn L G, Averitt R D and Nastasi M 2004 *J. Vac. Sci. Technol. A* **22** 2342
[14] Guruvenket S, Ghatak J, Satyam P V and Mohan Rao G 2005 *Thin Solid Films* **478** 256
[15] Morin P, Raymond G, Benoit D, Maury P and Beneyton R 2012 *Applied Surface Science* **260**

Journal of Physics: Conference Series **748** (2016) 012006 doi:10.1088/1742-6596/748/1/012006

69
[16] Kulczyk-Malecka J, Kellya P J, West G, Clarke G C B and Ridealgh J A 2014 *Surface & Coatings Technology* **255** 37

Deuterium removal from radiation damage in tungsten by isotopic exchange with hydrogen atomic beam

O V Ogorodnikova[1], S Markelj[2], V S Efimov[1] and Yu M Gasparyan[1]

[1]National Research Nuclear University MEPhI (Moscow Engineering Physics Institute), 31 Kashirskoe highway, Moscow 115409, Russia
[2]Jožef Stefan Institute and Association EURATOM-MHEST, 39 Jamova cesta, Ljubljana 1000, Slovenia

Corresponding author e-mail: olga@plasma.mephi.ru

Abstract. The tungsten samples were pre-irradiated with self-ions to create radiation-induced defects and then exposed to the deuterium atomic beam. The deuterium removal was studied by isotopic exchange with atomic hydrogen beam. Modification of the deuterium depth profile in self-ion irradiated tungsten under isotopic exchange up to a depth of 6 μm was measured *in-situ* by nuclear reaction analysis. The total deuterium retention after isotopic exchange was measured by thermal desorption spectroscopy. It is shown that the efficiency of the deuterium removal increases with increasing of the hydrogen incident flux, incident energy and temperature of the tungsten sample.

1. Introduction

Due to its physical properties such as low coefficient of physical sputtering and high melting temperature, tungsten (W) is considered as a priority plasma-facing material for fusion reactors [1-5]. In the burning fusion plasmas, plasma-facing components will be exposed to intense fluxes of deuterium (D), tritium (T) and helium (He) particles and neutrons (n) with energy of 14 MeV arising in the course of the D-T reaction. The decrease of tritium in fusion reactor materials is today one of the main requirements for materials that come into contact with the fusion plasma because of both the safety limitation of total tritium retention in fusion reactors and tritium self-sufficiency [3-5]. In order to stay below the safety limit, the fuel removal methods from the wall materials need to be developed. One possible method is the isotope exchange [6-8]. In the present paper, the isotope exchange technique is examined to be used as tritium removal method from radiation damage in W.

Tungsten (W) samples were first pre-irradiated with self-ions up to a fluence of $1.6\text{x}10^{18}$ W/m^2 to create radiation damage (0.5 dpa) [9] and then exposed to atomic D beam at sample temperature of either 600 K or 700 K. After it, the H atomic beam was used to remove D at the sample temperature of either 600 K or 500 K. The atomic H or D beam is created by thermal dissociation of H or D molecules in a hot capillary of the atomic beam source, HABS [10]. The capillary was heated with the same power of 195 W for all exposures providing capillary temperature of 2150 K. H_2 dissociation is typically from 80 to 98% depending on operation conditions. The source produces only neutral atoms without any high-energy particles and ions. The translation kinetic energy of hydrogen atoms can be estimated as $E_{kin}=(3/2)kT$ that is about 0.28 eV for the present operational temperature of the W capillary. The H_2 or D_2 gas pressure in the gas supply side was controlled by an all-metal leak valve and measured by an absolute capacitance manometer, Baratron by MKS. The average driving pressure during the atom exposures of tungsten sample was 250 ± 10 mbar for D_2 and 190±10 mbar for H_2. The distance between the sample surface and the capillary exit was 80 mm and the angle between the atom beam and the surface normal was 51°.

The atomic beam is distributed inhomogeneously over loaded surface with a maximum flux in the middle of sample. The maximum D and H atomic flux densities at the sample were 5.2×10^{18} D/m^{2}s and 6.7×10^{18} H/m^{2}s, respectively. The beam spot size and position was determined by observing the burn trace after short exposure of a millimetre paper mounted at the location of the sample. The average flux densities at the probing beam position were $(4.5\pm0.3)\times10^{18}$ D/m^{2}s for D atoms and $(5.6\pm0.1)\times10^{18}$ H/m^{2}s for H atoms [12]. Modification of the deuterium depth profile in self-ion irradiated tungsten during isotopic exchange up to a depth of 6 μm was measured *in-situ* by nuclear reaction analysis (NRA) using $^3He^+$ as the analysing beam as described in [11,12]. Kinetics of the D release from radiation-induced defects of specimens after isotopic exchange was studied by thermal desorption spectroscopy (TDS). The samples were heated resistively up to 1500 K with a linear ramp of 2 K/s.

2. Results and discussion

The D depth profile obtained after the D atomic exposure at sample temperature of 700 K up to a fluence of 7.8×10^{23} D/m^2 is shown in figure 1. Modification of the D depth profile after the H atomic beam loading at 500 K at different fluences is also shown in figure 1. More efficient isotopic exchange occurs near the surface up to 1 μm, however, the D concentration only slightly decreases after the H atomic flux of ~1×10^{23} H/m^2 and significant removal of the D up to 1 μm takes place at H atomic fluence of ~2×10^{24} H/m^2. The noticeable change of the D concentration at a peak damage was observed only at the highest fluence of the present experiment of 5.71×10^{24} H/m^2 (284.5 h). This is in an agreement with diffusion model with trapping of hydrogen at radiation defects reported in [13]. As it was found in [13], the D atoms reach the peak damage of ~1.3 μm at a fluence of about 6×10^{24} D/m^2 at 500 K. The isotopic exchange at the same depth also starts at ~5.7×10^{24} H/m^2 at 500 K as one can see in figure 1. This means that diffusion of H is the limiting process that defines the efficiency of isotopic exchange.

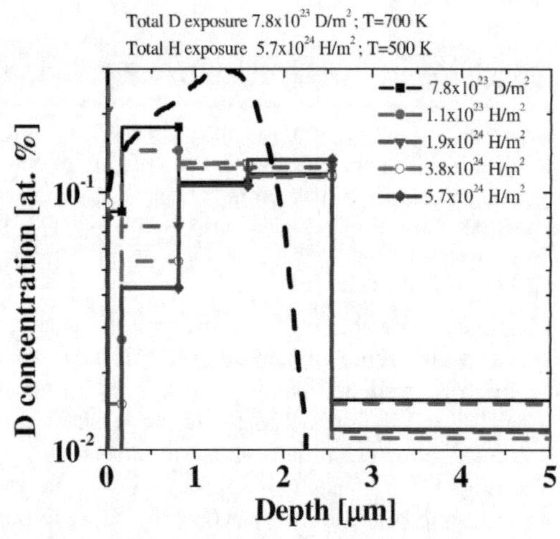

Figure 1. The D depth profile after the D atomic exposure at sample temperature of 700 K up to a fluence of ~7.8×10^{23} D/m^2 (48 h) and sequentially after the H atomic exposure at sample temperature of 500 K at different fluences up to a fluence of ~5.7×10^{24} H/m^2 (284.5 h). Dashed line shows the damage profile calculated by SRIM [9]. The D atomic flux density was 4.5×10^{18} D/m^{2}s and H atomic flux density was 5.6×10^{18} H/m^{2}s.

After the isotope exchange of D by H at 500 K and NRA measurements of the D concentration up to 6 μm, total D concentration in W was measured by TDS. TDS analysis has been performed several months after atomic exposure. A quadrupole mass spectrometer (QMS) was used to monitor the desorbed hydrogen molecules: mass 2(H_2), mass 3 (HD) and mass 4 (D_2), and D-related water: mass 19 (HDO) and mass 20 (D_2O). The absolute sensitivity of QMS to D_2 was routinely measured after every experiment using a system including two leak valves and a baratron (see details in [14]). The relative sensitivity to H_2 was determined once in a separate experiment with the same procedure. The relative sensitivity for HD was taken as an average between H_2 and D_2 values. Figure 2 shows total D release from self-ion damaged W after the only D exposure at 700 K up to a fluence of 3.78×10^{23} D/m^2 (see details in [13,15]) and after the isotopic exchange experiment. Total D retention was defined as 2·signal(D_2)+signal(HD)/2. TDS of D from the sample without H exposure shows higher D retention (6.4×10^{19} D/m^2) compared to D in the sample after H exposure (4.7×10^{19} D/m^2) (figure 2). The total D retention measured by TDS many months after atomic exposure is much lower compared to the D retention measured by NRA up to 6 μm. TDS of D from sample without H exposure is narrower than that from sample after isotopic exchange. The comparison of release of D in the form of D_2 and HD is shown in figure 3. For the experiment of only D atomic beam exposure, mass 4 (D_2) was dominating the release spectrum. Obviously, more D is desorbed in the form of HD (mass 3) than in the form of D_2 (mass 4) after H exposure as shown in figure 3 because of high density of hydrogen in the sample.

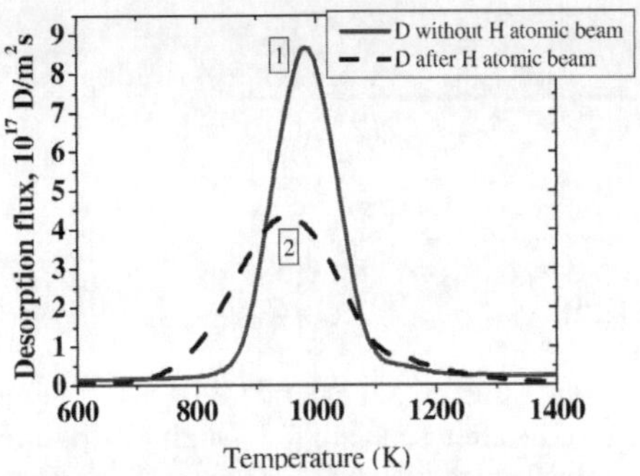

Figure 2. Thermal desorption spectra of total D release from self-ion irradiated W (0.5 dpa) after exposure to the D atomic beam at 700 K up to a fluence of 3.8×10^{23} D/m^2 (curve 1) and after exposure to D atomic beam at 700 K up to a fluence of 7.8×10^{23} D/m^2 with sequential exposure to H atomic beam at 500 K up to a fluence of ~5.7×10^{24} H/m^2 (284.5 h). Total D was defined as 2·signal(D2)+0.5·signal (HD).

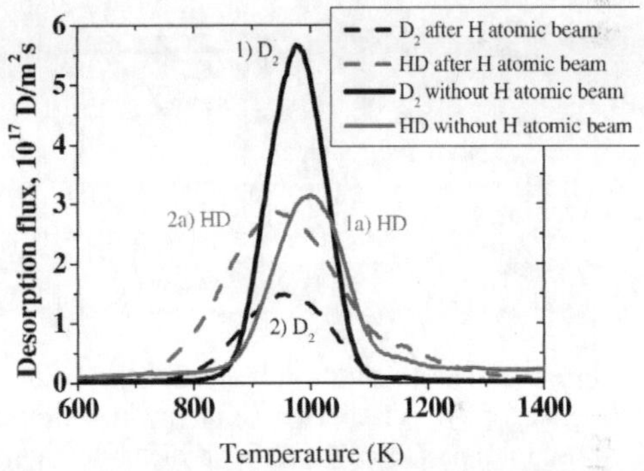

Figure 3. Thermal desorption spectra of D from self-ion irradiated W (0.5 dpa) after exposure to the D atomic beam at 700 K up to a fluence of 3.8×10^{23} D/m^2 (curve 1 is for the D released as mass 4 and curve 1a is for D released as mass 3) and after exposure to D atomic beam at 700 K up to a fluence of 7.8×10^{23} D/m2 with sequential exposure to H atomic beam at 500 K up to a fluence of ~5.7×10^{24} H/m2 (284.5 h) (curve 2 is for the D released as mass 4 and curve 2a is for D released as mass 3).

An increase of the sample temperature results in an increase of diffusion of hydrogen isotopes [13] and, consequently, the rate of replacing of D atoms by H atoms increases. Therefore, the efficiency of isotopic exchange increases. Figure 4 shows the D depth profile after the D atomic exposure at sample temperature of 600 K and modification of this profile after sequential H atomic exposure at the same temperature. Between the end of the D exposure and start of H exposure the sample was kept at 600 K during 20 hours to study the effect of the D release under the annealing in vacuum and exclude the impact of this effect on the isotopic exchange. The annealing in a vacuum only slightly reduces the D concentration at a peak damage (~1.3 μm) by a factor of ~1.5. While the isotopic exchange during 20 hours reduces the D concentration at a peak damage by a factor of ~3.3. Therefore, the efficiency of isotopic exchange for D removal from self-ion irradiated W is 2 times higher than annealing in vacuum as it was also stated in [11]. The increase of the D release from radiation-induced defects in W by exposure to H plasma and by heating in H_2 atmosphere compared to annealing in vacuum was also reported in [16,17] and [18], respectively. Increasing of the H fluence results in further removal of D from damaged zone by replacing with H and diffusion of D both to the surface to be desorbed and into the bulk to be re-trapped. At H atomic fluence of ~$2x10^{24}$ H/m^2, the D is removed not only from radiation-induced defects in damaged zone but also from bulk intrinsic defects which are out of the damaged zone.

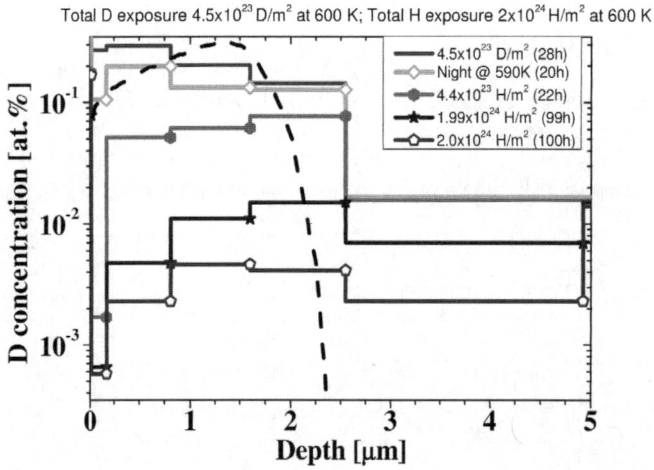

Figure 4. The D depth profile after the D atomic exposure at sample temperature of 600 K up to a fluence of ~$4.3x10^{23}$ D/m^2 and modification of this profile after sequential H atomic exposure at sample temperature of 600 K at different fluences up to a fluence of ~$2x10^{24}$ H/m^2 (100 h). Dashed line shows the damage profile calculated by SRIM [9]. The H atomic flux is ~$5.6x10^{18}$ H/m^{2}s.

According to the analytical expression derived in [14], the time to decorate radiation-induced defects up to a depth x decreases with ion flux and ion energy

$$t = (x^2 / 2I_0 R_p) \sum_{k=1}^{m} W_{tk}(T), \tag{1}$$

where I_0 is the ion flux, ξ is the penetration probability, R_p is the mean ion range and $W_t = \sum_{k=1}^{m} W_{tk}$ is the total density of defects (m is the number of defects), which can be occupied by D at certain temperature T. The D diffuses deeper at higher temperatures or in the case of small trap density. In the case of the exposure of W to D atoms at 700 K, only the defects with high binding energy (deep traps) are populated with D. Whereas H populates both kinds of defects with low and high binding energies under exposure of W to H atoms at 500 K. Because the rate of D and H diffusion is different only by a factor of $\sqrt{2}$ at the same sample temperature [19], the fact that the isotopic exchange at certain depth x starts approximately at the same fluence when the H atom reaches this depth confirms that H atom re-places D atom from deep traps. In the case of the exposure of W to D atomic beam at 600 K and sequentially to H atomic beam at 600 K, the D from both shallow and deep defects is exchanged with H. The rate of this exchange depends mainly on the diffusion of H atoms, namely, on the sample temperature in the present experiments, but it seems that this rate does not depend on the initial concentration of deuterium. To confirm or reject this suggestion and define probabilities of the D de-population from deep and shallow defects by exchange with H, we will populate only deep defects with D using the D atomic exposure at 700 K (in contrast with the present experiment shown in figure 4 where we populated with D both kinds of defects with low and high binding energies using the D atomic exposure at 600 K) and sequential exposure to H atoms at 600 K in the future experiment. The efficiency of isotopic exchange increases with increasing the sample temperature, incident ion/atomic flux and incident H energy, as it was derived analytically in [19], that is consistent with the present experimental data and experimental data at different ion fluxes and energies reported in [16-18].

Acknowledgements
This work is supported by Russian Science Foundation (RSF) grant №16-12-10332 and by National Research Nuclear University MEPhI in the framework of the Russian Academic Excellence Project (contract No. 02.a03.21.0005, 27.08.2013).

References

[1] Federici G, Skinner C H, Brooks J N et al 2001 *Nucl. Fusion* **42** 1967
[2] Tobita K, Nishio S, Enoeda M et al 2006 *Fusion Eng. Des.* **81** 1151
[3] Pitts R A, Carpentier S, Escourbiac F et al 2013 *J. Nucl. Mater.* **438** S48
[4] Loarer T et al 2013 *J. Nucl. Mater.* **438** S108
[5] Roth J, Tsitrone E, Loarer T, Philipps V, Brezinsek S et al 2008 *Plasma Phys. Control. Fusion* **50** 103001
[6] Watanabe T, Kaneko T, Matsunami N, Ohno N, Kajita S, Kuwabara T 2015 *J. Nucl. Mater.* **463** 1049, http://dx.doi.org/10.1016/j.jnucmat.2014.12.011
[7] Takagi I, Nomura S, Minamimoto T, Akiyoshi M, Kobayashi T, Sasaki T 2015 *J. Nucl. Mater.* **463** 1125, http://dx.doi.org/10.1016/j.jnucmat.2014.11.037
[8] Loarer T, Brezinsek S, Philipps V, Romanelli-Gruenhagen S, Alves D, Carvalho et al 2014. *J. Nucl. Mater.* **463** 1117 doi:10.1016/j.jnucmat 11.012
[9] Ogorodnikova O V and Gann V 2015 *J. Nucl. Mater.* **460** 60
[10] Tschersich K G, Fleischhauer J P, Schuler H 2008 *J. Appl. Phys.* **104** 034908
[11] http://www.mbe-components.com/products/gas/habs.html
[12] Markelj S et al 2016 *J. Nucl. Mater.* **469** 133
[13] Markelj S, Ogorodnikova O V, Pelicon P, Schwarz-Selinger T, Vavpetic P, Cadez I 2014 *Phys. Scr.* T159 014047
[14] Ogorodnikova O V, Markelj S, von Toussant U 2016 *J. Appl. Phys.* **119** 054901
[15] Rusinov A, Gasparyan Yu, Perelygin S, Pisarev A, Stepanov S, Trifonov N 2009 *Instr. Exp. Tech.* **52** 871
[16] Gasparyan Yu M, Ogorodnikova O V, Efimov V S, Mednikov A, Marenkov E D, Pisarev A A, Markelj S, Čadež I 2015 *J. Nucl. Mater.* **463** 1013
[17] Barton J L et al. 2013 *J. Nucl. Mater.* **438** S1183
[18] Alimov V Kh et al. 2011 *Physica Scripta* **T145** 014039
[19] Hatano Y et al. 2013 *J. Nucl. Mater.* **438** S114
[20] Ogorodnikova O V 2015 *J. Appl. Phys.* **118** 074902

Journal of Physics: Conference Series **748** (2016) 012008 doi:10.1088/1742-6596/748/1/012008

Deuterium retention after deuterium plasma implantation in tungsten pre-damaged by fast C^+ ions

V S Efimov[1], Yu M Gasparyan[1], A A Pisarev[1], B I Khripunov[2], V S Koidan[2], A I Ryazanov[1,2] and E V Semenov[2]

[1]National Research Nuclear University MEPhI (Moscow Engineering Physics Institute), 31 Kashirskoe highway, Moscow 115409, Russia
[2]NRC "Kurchatov Institute", 1/1 Ac. Kurcharov sq., Moscow 123182, Russia

Corresponding author e-mail: evitaliy2005@yandex.ru

Abstract. Thermal desorption of deuterium from W was investigated. Virgin samples and samples damaged by 10 MeV C^{3+} ions were implanted from plasma in the LENTA facility at 370 K and 773 K. In comparison with the undamaged sample, deuterium retention in the damaged sample slightly increased in the case of deuterium implantation at RT, but decreased in the case of deuterium implantation at 773 K. At 773 K, deuterium was concluded to diffuse far behind the D ion range in the virgin sample, while C implantation region was concluded to be a barrier for D diffusion in the damaged sample.

1. Introduction

Tungsten is considered as a plasma facing material in future fusion reactors, where it will be exposed to extreme heat and particle fluxes from plasma and to high energy neutrons and α particles generated in D-T reaction. Radiation damage in tungsten will lead to degradation of mechanical properties, and also it may increase hydrogen isotope retention. Irradiation by fast ions is often used for simulation of neutron impact on materials. In a series of works [1,2], 10 MeV C^{3+} ions were used for damage production and serious changes in material were observed after irradiation. In particular, swelling of the damaged layer was reported in [1]. Deuterium accumulation in damaged tungsten after D plasma exposure was analyzed in those experiments only in a very thin surface layer using ERD (Elastic Recoil Detection) technique.

In this work, deuterium retention after plasma exposure was measured by thermal desorption spectroscopy (TDS) in the samples identical to those used previously in [1]. TDS provides information about the total deuterium retention in sample.

2. Experimental

Polycrystalline tungsten samples with the purity of 99.95 % were used in experiments. Some samples was preliminary damaged by 10 MeV C^{3+} ions with the fluence of $(1.5\text{-}2)\times10^{21}$ m^{-2}. According to SRIM calculations, this irradiation produces the average damage of 6.3-8.4 dpa with the maximum of 30-40 dpa at 3.5 μm [1]. The average concentration of carbon atoms near the surface was less than 0.01 at.%, while maximal concentration of carbon atoms reached several at.% at 3.5 μm [1].

Two damaged and two virgin undamaged tungsten samples were exposed to deuterium plasma in the LENTA facility (Kurchatov Institute) in pair at room temperature (< 370 K) and at 773 K. The ion flux in the continued operation regime was (1.3-2)$\times10^{21}$ D/m^{2}s, and the fluence was of about 10^{25}-10^{26} D/m^2. The energy of incident ions was 250 eV.

Deuterium retention in the samples after plasma exposure was investigated by means of thermal desorption spectroscopy (TDS) using the UHV TDS stand (MEPhI). The base pressure in the TDS chamber was less than 5×10^{-9} mbar. The samples were introduced into TDS chamber through the lock chamber with a separate pumping system that gives a low background signal during TDS runs. The sample temperature was measured by the W–Re thermocouple directly welded to the edge of the sample. The heating rate was 2 K/s, and the maximal temperature was 1450 K. Desorption of deuterium containing molecules (D_2, HD, HDO, D_2O) was measured by a quadrupole mass spectrometer (QMS). Calibration of QMS measurements was done after each experiment. Description of the TDS-stand and experimental procedure are given in [3] in more details. Table 1 summarizes the information about the investigated samples. Letters v and d denote virgin and damaged samples, respectively, while the numbers are the temperatures during deuterium implantation.

Table 1. Description of the samples used.

Sample	C^{3+} ions fluence, m^{-2}	Primary defect concentration (av. in 3.5 μm), dpa	Temperature, K	Plasma exposure fluence, D/m^2	Erosion depth, μm
v370	0	0	<370 (RT)	5.8×10^{25}	2.3
v773	0	0	773	1×10^{25}	2.2
d370	2×10^{21}	8.4	<370 (RT)	3.0×10^{25}	1.3
d773	1.5×10^{21}	6.3	773	1.2×10^{25}	0.8

3. Experimental results and discussion

Deuterium thermal desorption spectra for the four samples are given in figure 1. The total deuterium retention for all samples is compared in figure 2. The data for the damaged sample are multiplied by 10 to be plotted in figure 1.

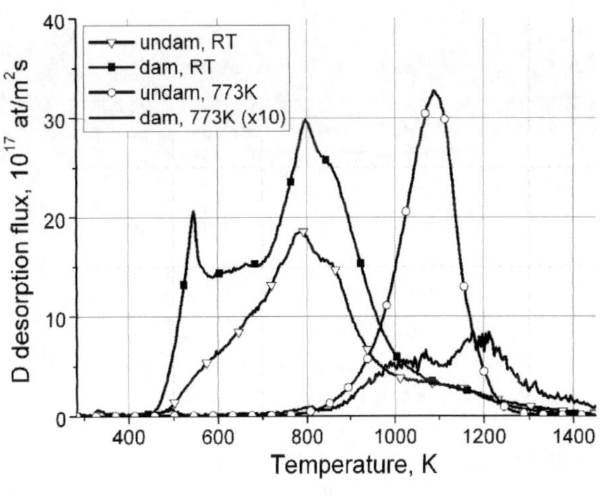

Figure 1. Deuterium thermal desorption spectra (D_2) for damaged and virgin W samples irradiated in deuterium plasma.

In the case of implantation at room temperature, deuterium desorption takes place in a wide temperature range from 450 K to 1400 K with several peaks. The maximum desorption flux is at 800 K for both (damaged and virgin) samples, but the number of deuterium is slightly higher in the damaged sample. In the case of implantation at 773 K, the desorption starts at the temperature of implantation. For the virgin sample, the spectrum has one pronounced desorption peak with the maximum at 1100 K. For the damaged sample, the spectrum has a maximum at about the same temperature of 1050 K and an additional maximum at 1200 K. The striking effect is that the number of deuterium in the damaged sample is significantly less than that in the virgin sample.

Deuterium after plasma implantations is retained mainly due to trapping in intrinsic defects and defects produced by deuterium implantation and radiation damage. Diffusion of deuterium into the W bulk in presence of both types of defects is low at low temperatures, so one can suggest that deuterium is trapped mainly in a thin near surface layer below the carbon ion range. TDS from the virgin and damaged samples for low temperature implantation take place nearly in the same temperature range. This can mean that intrinsic and radiation defects in W are of the same type, and the difference is only in the quantity of defects. The deuterium retention for low temperature plasma implantation is nearly the same for the virgin and damaged samples, and it is only slightly higher for the damaged sample. This means that the net concentration of radiation defects is less than that of intrinsic defects in the region of trapping at low temperature. One can conclude from this that the trapping takes place very close to the surface well below the C ion range, where the concentration of radiation defects is very small as it was discussed above. The obvious difference between the amplitudes of two peaks in the spectra for the damaged and virgin samples is due to difference in the number of defects of different binding energies in the virgin and damaged samples.

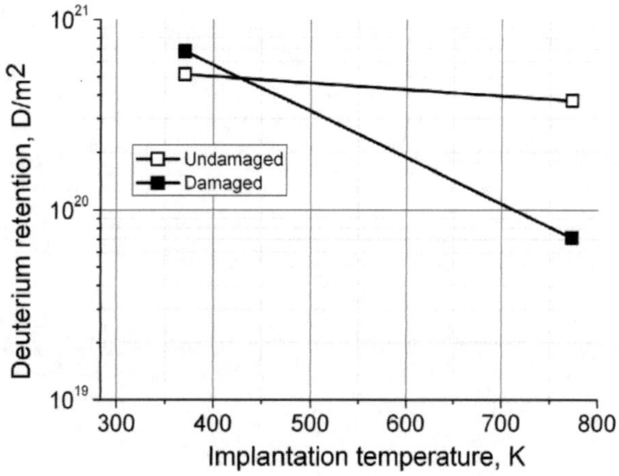

Figure 2. Total deuterium retention in four samples.

Defect trapping usually decreases with temperature, and one can see that this is the case also in these experiments both for the virgin and damaged samples. At the elevated temperature of implantation, defects with a low binding energy do not retain deuterium during implantation. This is why TDS after implantation at 773 K have only high temperature parts that start from the temperature of implantation.

The prominent effect for the virgin sample is that the high temperature part of TDS after irradiation at 773 is more intensive than after irradiation at RT. The high temperature peaks in TDS can be explained either by defects of higher binding energy or by trapping in and release from deeper layers of the material, which are far from the surface. If to accept that the defect structure in the virgin sample does not change significantly in the temperature range below 773 K, one may suggest that the high temperature peak for the virgin sample implanted at 773 K is due to intrinsic defects that are deeper in the sample bulk. These defects are populated due to deuterium diffusion. With increase of the temperature, diffusion rate increases, traps far from the surface populate, and the high temperature part of TDS increases. This is what one can see in figure 1 for the virgin sample. Moreover the total amount of deuterium trapped in the virgin sample only slightly decreases with temperature. Diffusion into the bulk is suggested also in [4, 5] and some other experiments.

The high temperature peaks for the damaged sample seem to be also due to defects placed far from the surface outside the D implantation range, but increase of temperature leads to decrease of the high temperature TDS peaks and drastic decrease of the total number of the deuterium trapped in the damaged sample. This can be explained if deuterium diffusion into the bulk at high temperatures is blocked in some way after irradiation by carbon ions. The opposite and expectable result was observed in [4,5]: retention in a W sample damaged by MeV tungsten ions increased, though the damage level in our experiment was remarkably higher than in [5]. The difference between [5] and this work could be due to effect connected to implanted carbon atoms, which may be the reason of the diffusion barrier for deuterium atoms. If this barrier is connected with C irradiation, it must be placed at the implantation range of carbon, which is behind the implantation range of deuterium. Due to this barrier, deuterium can not diffuse deep into the bulk even at 773 K, and both the high temperature peaks and total number of deuterium in TDS decrease due to irradiation by high energy carbon ions. According to calculations, the main part of carbon atoms was concentrated at the depth of 3.5 μm, while the concentration near the surface was less than 0.01 at.%. The consideration given above leads to the conclusion that the high temperature peaks correspond to deuterium trapped at depths much larger than the carbon ion range. The retained fluences of deuterium in the virgin and damaged samples implanted at 773 are similar, but the depth distributions are very different. The virgin sample has a rather high D concentration at small depths close to the D implantation range, while the damaged sample has a very small D concentration with a wide concentration profile spreading well behind the range of C ions.

The barrier effect of carbon and nitrogen was proposed also for explanation of experiments on ion-driven deuterium permeation [6] and deuterium accumulation [7] in tungsten: it was observed that both permeation and accumulation increased if impurities (C, N) were added to the incident flux. This was explained by formation of a thin barrier layer on the surface, which prevented re-emission of implanted deuterium back in vacuum.

4. Summary

Deuterium trapping in tungsten damaged by 10 MeV C^{3+} ions was investigated by means of thermal desorption spectroscopy. The damage led to some increase of D trapping at 370 K but drastic decrease of deuterium trapping at 773 K in comparison with the virgin sample. The experimental data were interpreted in the following way:

Trapping is due to both intrinsic and radiation induced defects.

At 370 K deuterium diffusion is slow in the field of defects and deuterium is trapped to a high concentration in the near surface region not far from the D implantation range.

At 773 K deuterium diffusivity is high enough to transport D atoms into the W bulk. Therefore deuterium diffuses far from the surface, its concentration is very small but the diffusion length is high, so the integral trapping is about as high as at 370 K implantation.

Carbon irradiation produces a diffusion barrier behind the D implantation range that prevents D diffusion into the W bulk. Therefore in the case of D implantation at 773 K, the number of trapped deuterium drastically decreases in the sample pre-irradiated by C ions.

Acknowledgements

This work was supported in part by National Research Nuclear University MEPhI in the framework of the Russian Academic Excellence Project (contract No. 02.a03.21.0005, 27.08.2013).

References

[1] Khripunov B I, Koidan V S, Ryazanov A I et al 2015 *Physics Procedia* **71** 63
[2] Khripunov B, Gureev V, Koidan V et al 2015 *J. Nucl. Mater.* **463** 258
[3] Rusinov A, Gasparyan Yu, Perelygin S et al., *Instrum. Exp. Tech.* 52 (2009) 871

[4] Gasparyan Yu M, Ogorodnikova O V, Efimov V S, et al 2015 *J. Nucl. Mater.* **463** 1013
[5] Ogorodnikova O, Sugiyama K 2013 *J. Nucl. Mater.* **442** 518
[6] Peng H Y, Lee H T, Ohtsuka Y, Ueda Y 2011 *Phys. Scr* **T145** 014046
[7] Ogorodnikova O V, Sugiyama K, Markin A, et al 2011 *Phys. Scr.* **T145** 014034

Thermal cycling and high power density hydrogen ion beam irradiation of tungsten layers on tungsten substrate

A A Airapetov, L B Begrambekov, I Yu Gretskaya, A V Grunin, M Yu Dyachenko, N A Puntakov and Ya A Sadovskiy

National Research Nuclear University MEPhI (Moscow Engineering Physics Institute), 31 Kashirskoe highway, Moscow 115409, Russia

Corresponding author e-mail: grunin@plasma.mephi.ru

Abstract. Tungsten layers with iron impurity were deposited on tungsten substrates modeling re-deposited layers in a fusion device. The samples were tested by thermocycling and hydrogen ion beam tests. Thermocycling revealed globule formation on the surface. The size of the globules depended on iron impurity content in the coating deposited. Pore formation was observed which in some cases lead to exfoliation of the coatings. Hydrogen ion irradiation lead to formation of blisters on the coating and finally its exfoliation.

1. Introduction

Tungsten is considered as a plasma-facing material of the ITER divertor. The behavior of tungsten during thermal cycling and under plasma radiation was researched in a number of articles [1-3]. Tungsten atoms sputtered by plasma radiation form layers of tungsten on divertor tiles less affected by irradiation [4, 5]. This work is dedicated to researching the processes in the deposited layers during thermal cycling and high power density hydrogen ion beam irradiation.

2. Experimental

Tungsten layer deposition and ion flux irradiation was conducted on the CODMATT (COating Deposition and MATerial Testing) stand [6]. The layers were formed on polished tungsten plates with the dimensions of $15\times7.5\times1$ mm^3 using atoms sputtered from the tungsten target by plasma ions. The layers of 2 μm thickness were formed with the deposition rate of 2 μm/hour.

It should be noticed that while researching the behavior of tungsten re-deposited in the ITER conditions, beryllium and, possibly, stainless steel impurities sputtered from different parts of the installation will be present in the layers. As a safety precaution, beryllium cannot be used in our laboratory, which is why stainless steel (Fe, Cr, Ni) atoms were directed towards the tungsten surface to be used as an impurity.

Thermal cycling of samples was conducted in the Multifunctional Research Complex of Mass-spectrometry Analysis (MIKMA) [7]. The samples were heated by heat radiation from the tungsten coil located at the opposite side of the sample and heated by applied current in a vacuum of at least 10^{-6} Torr. The upper limit of the temperature was 1200 °C. The temperature gradient on the sample did not exceed ±10°C. The samples were heated at the rate of 5°C/sec and cooled for 5 minutes down to 300 °C. A number of samples with a small (0.2% at.) and large (2% at) concentration of iron impurity in the deposited layer has been prepared for testing.

The deposited layers' structure and texture analysis before and after the tests was conducted on the Vega Tescan 3 scanning electron microscope (SEM), the cross-section was prepared using the FEI Quanta 200 3D system.

3. Results

3.1. Thermal cycling of the deposited layers

After 260 cycles signs of restructuring have begun appearing on the surface of low impurity level deposited film. Distinct globules 200-500 nm in diameter have appeared (figure1). No signs of cracking or exfoliation were found.

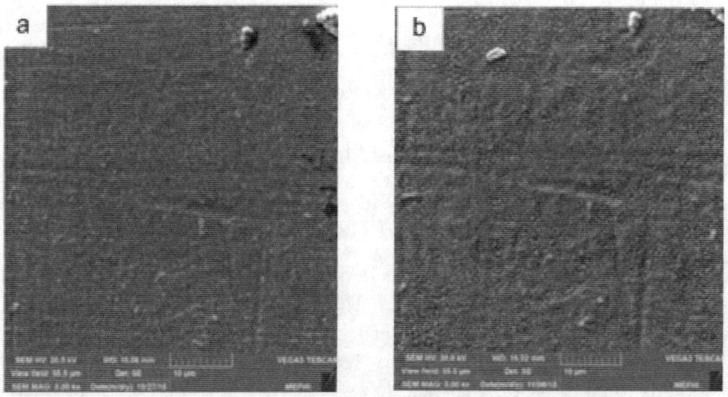

Figure 1. Deposited tungsten with small impurity concentration. a) directly after deposition, b) after 260 heating cycles.

The same experiments were carried on the tungsten layer with a large iron concentration (2% at.). Thermal cycling has lead to the formation of larger globules (1-3 μm in diameter, figure 2), which may be connected with the crystallization processes in the deposited layer. After 250 cycles the formation of pores on the surface is clearly visible. After 350 cycles the pore formation on the surface continues, with signs of exfoliation of the deposited layer in some areas, which may hint towards the formation of a large number of pores in the bulk of the deposited tungsten.

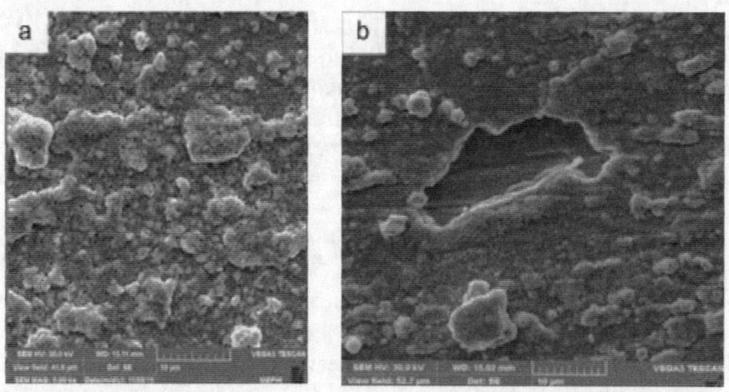

Figure 2. The deposited layer of tungsten with 2% at iron impurity content a) pores on the surface after 250 cycles, b) exfoliation after 350 heating cycles.

Journal of Physics: Conference Series **748** (2016) 012009 doi:10.1088/1742-6596/748/1/012009

3.2. Hydrogen ion irradiation of the deposited layers

During the sample irradiation of tungsten film with 2% at iron impurity concentration with 5 Mw/m^2 power density H_2^+ ions (ion energy is 10 KeV, cyclic irradiation mode with 20 pulses of 5 seconds each), blister formation was observed on the surface of the layer, with the temperature of the surface not exceeding 1200°C, with the sample cooled down to 600°C between the pulses. Fig. 3a shows the image of the film after the irradiation. Blister formation was not observed on the film with low impurity level irradiated under the same conditions. Formation of a large number of microscopic gas bubbles near the film-substrate boundary can be seen on figure 3b.

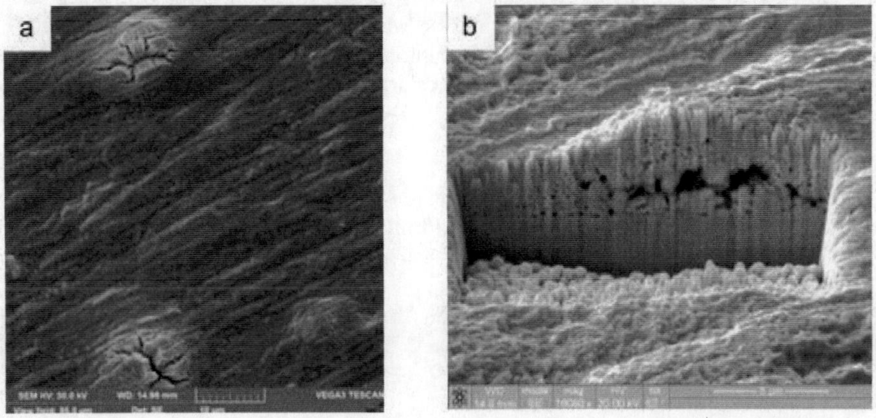

Figure 3. Deposited layers of tungsten after hydrogen ion beam irradiation. a) distinct blisters, b) blister cross-section.

In the figure 4 one could observe the stages of blister formation under hydrogen ion irradiation.

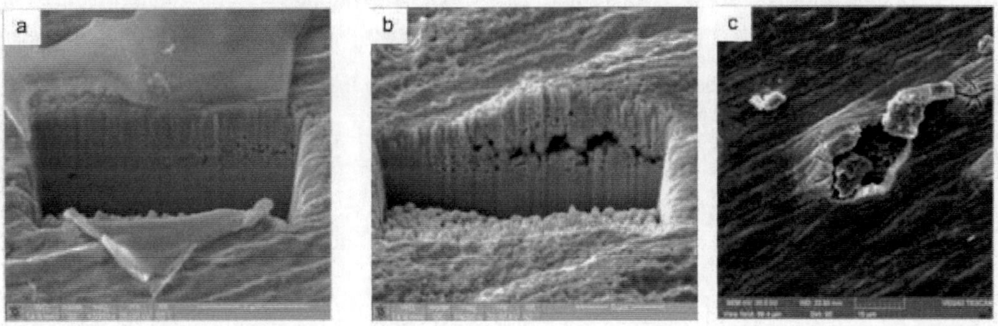

Figure 4. The stages of blister formation: a) gas bubbles formation, b) growth and association of adjacent bubbles, c) destruction of blister.

On the first stage the gas bubbles near the separation line between deposited layer and the substrate are formed (figure 4a). Then the bubbles growth and association lead to the occurrence of stresses in deposited layer and as a consequence blister formation (figure 4b). Further hydrogen accumulation leads to the blister cover destruction.

Hydrogen accumulation in these bubbles could lead to formation of tensions in the deposited layer parallel to the surface and the subsequent blister formation. Interesting thing to note is that open pores were formed when the sample was heated to the same temperature without irradiation which could assist with gas captured during the deposition escaping the sample and, as a result, the blisters and tensions did not form.

4. Conclusions

Thermal cycling and hydrogen ion beam irradiation of samples of tungsten film on tungsten with different concentrations of iron impurity in the deposited layer (0,2% at. and 2% at.) were conducted. During the thermal cycling, globules were formed on the surface of the films, with larger globules appearing on the sample with higher impurity concentration, with the diameter of 1-3 μm. Pore formation was also observed on the samples with higher impurity concentration, which lead to the exfoliation of the deposited layer. This phenomenon was not observed on the sample with 0,2% at. iron impurity concentration. During the hydrogen ion beam irradiation of the samples the pore development on the samples with high impurity concentration lead to exfoliation of the deposited layer and formation of blisters.

Acknowledgments

This research was completed using the equipment of the CCU IC RAS, with the support of the Ministry of Education and Science (project RFMEFI62114X0005) and of National Research Nuclear University MEPhI in the framework of the Russian Academic Excellence Project (contract No. 02.a03.21.0005, 27.08.2013).

References

[1] Krieger K, Geier A, Gong X, Maier H, Neu R, Rohde V 2003 *J. Nucl. Mater.* **313–316** 327
[2] Buzi L, Temmerman G, Unterberg B, Reinhart M, Litnovsky A, Philipps V, Van Oost G , Sören Möller 2014 *J. Nucl. Mater.* **455** 316
[3] Krasheninnikov S I 2011 *Plasma Phys. Control. Fusion* **53**(8) 083001
[4] Balden M, Rohde V, Lindig S, Manhard A, Krieger K 2013 *J. Nucl. Mater.* **438** S220
[5] Krat S, Coad J P, Gasparyan Yu, Hakola A, Likonen J, Mayer M, Pisarev A, Widdowson A 2013 *J. Nucl. Mater.* **438** S742
[6] Azizov E, Barsuk V, Begrambekov L, Buzhinsky O, Evsin A, Gordeev A, Grunin A, Klimov N, Kurnaev V, Mazul I, Otroshchenko V, Putric A, Sadovskiy Ya, Shigin P, Vergazov S, Zakharov A 2015 *J. Nucl. Mater.* **463** 792
[7] Airapetov A, Begrambekov L, Brémond S, Douai D, Kuzmin A, Sadovskiy Ya, Shigin P, Vergasov S 2011 *J. Nucl. Mater.* **415** S1042

Simulation of diffusion of hydrogen atoms in the lattice of tungsten

N N Degtyarenko and A A Pisarev

National Research Nuclear University MEPhI (Moscow Engineering Physics Institute), 31 Kashirskoe highway, Moscow 115409, Russia

Corresponding author e-mail: eugen_mazur@mail.ru

Abstract. Results of DFT calculations of the activation energy for diffusion and vibrational spectra of interstitial hydrogen atoms in the lattice of tungsten are presented. The temperature dependences of the diffusion coefficient are calculated and compared with experimental data.

1. Introduction

For a long time, diffusion of hydrogen in tungsten has attracted interest for thermonuclear fusion. Early experimental works [1-5] were based on measurements of permeability and desorption and were carried out at high temperatures from 910 to 2500 K. The values of activation energies for diffusion obtained in those works were rather high from 0.86 to 1.8 eV; and only in the work of Frauenfelder [1] performed in the range of 1200-2000 K, a relatively low value of 0.4 eV was obtained. Experiments at low temperatures from 293 to 673 K were performed in later works [6-8] based on measurements of concentration profiles of radioactive tritium, and the activation energy for diffusion was 0.39 eV. Theoretical calculations based on first principles and performed at 0 K gave smaller values of the activation energy 0.19 eV [9] and 0.21 eV [10]. In a recent work [11], based also on first principles, two paths of transition between two interstitial tetrahedral sites were considered: a short jump between two closest tetrahedral sites, and a longer jump through an octahedral site. At the temperature of 0 K, the short jump had the activation energy of 0.22 eV, and the longer jump had the activation energy of 0.38 eV. To analyze the situation at higher temperatures, vibrations of tungsten atoms and thermal expansion of the lattice were taken into account. It was shown that activation energies for jumps increased from 0.18 eV to 0.40 eV for the short T-T jump and from 0.29 eV to 0.69 eV for the long T-O-T jump with increase of temperature from 300 K to 2500 K. Thus, the uncertainty in the activation energy for diffusion of hydrogen in tungsten remains rather large. This work is devoted to re-analysis of two paths of elementary jumps at 0 K and elevated temperatures basing on the DFT theory.

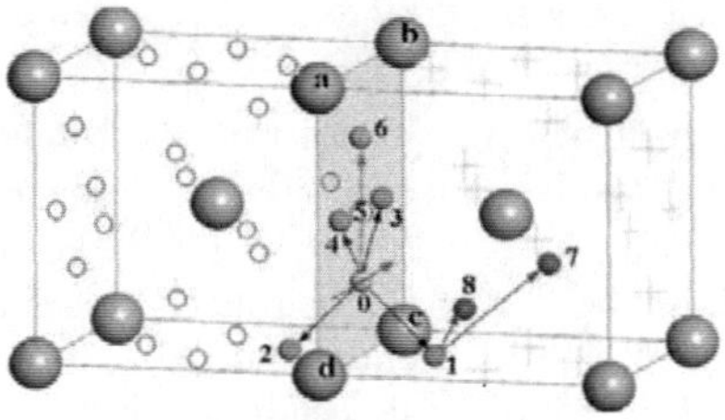

Figure1. Several positions of interstitial H atoms in the W unit cell. Positions 1-4 and 6-8 are tetrapores, position 5 is an octapore. Some short and long jumps between tetrahedral sites are shown.

Method

Optimization of the configuration of the system "W crystal – H atom" and calculations of energies were performed in the DFT framework using super cells with periodic boundary conditions. In calculations, GGA approximation with PBE exchange-correlation functional and pseudopotential that keeps the norm was used. Cut-off of the kinetic energy was at ~720 eV. Positions of all atoms of the super cell were optimized to minimize the total energy of the cell. Self-consistency threshold of energy convergence was 5×10^{-6} eV/atom during optimization of the system configurations.

In equilibrium, the hydrogen atom in the lattice of tungsten occupies a tetrapore, and figure 1 shows a few of them (positions 0-4 and 6-8). Also an octapore (position 5), where the energy of H atom is higher, is shown. Hydrogen atom can jump between two tetrapores in two ways: either directly between two nearest tetrapores (0-1 or 0-3, for example) or through an octapore (0-5-6, for example). Taking into account relaxation of the cell, the short distances 0-1, 0-2, 0-3, etc. are $\lambda_1 \approx a/\sqrt{8} \approx 1.13$ Å, and the long distances between 0-6, etc are $\lambda_2 \approx a/2 \approx$ of 1.6 Å. A hydrogen atom in position 0 can make 4 short jumps into sites 1-4, and 2 long jumps in sites 6 and 6* (symmetrically below position 0).

2. Energies and vibration frequencies at 0 K

The energy barriers for short and long jumps at 0 K are shown in figure 2. Calculations were performed using the linear and quadratic synchronous transition (LST/QST) with optimization. The short jumps 0-1, 0-3. etc need a smaller activation energy $\varepsilon_1 \approx 0.195$ eV, while the long jump 0-5-6 needs a higher activation energy $\varepsilon_2 \approx 0.37$ eV, which is almost equal to the value previously calculated in [12].

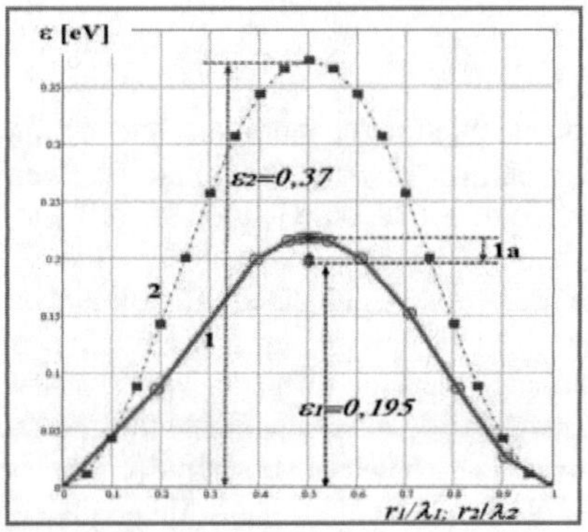

Figure 2. Total energy E_F along LST trajectories between two tetrapores: 1 – short path; 2 – long path, 1a – the energy within QST (one point at the top of the barrier).

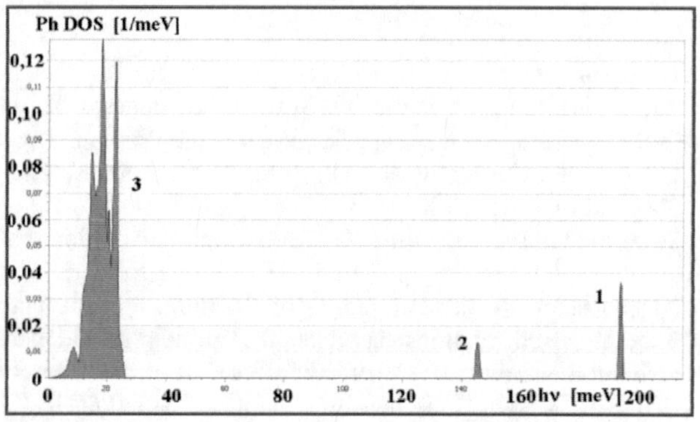

Figure 3. The density of phonon states (PhDOS) for W super cells with one interstitial H atom in an equilibrium tetrahedral site.

The phonon density of states (PhDOS) for W super cell with one H atom in the equilibrium interstitial tetrahedral site has several maxima (figure 3). Peak 1 is the oscillation of an atom in H modes with almost identical frequencies and amplitudes $h\nu_1 \approx 194$ meV (46.92 THz) and $h\nu_2 \approx 193.5$ meV (46.77 THz). The H atom oscillates along the line connecting nodes 0 and 6 that corresponds to the long jump through octapore. Peak 2 is the local vibration mode of the hydrogen atom with $h\nu_3 \approx 146$ meV (35.39 THz); and the direction of oscillations is perpendicular to the line 0-6 and coincides with none of short jump directions. There is also a set of oscillation modes of W atoms (group of peaks 3) in different directions with average frequency $h\nu_W \approx 16$ meV.

Calculations of PhDOS for the W supercell with one H atom on the top of the potential barrier for a short jump (0→1, e.g.) demonstrate three peaks associated with H vibration. Two local vibrations have real frequencies $h\nu^*_2 \approx 200$ meV (53.2 THz) and $h\nu^*_3 \approx 257$ meV (68.4 THz). There is also a mode of

H vibration with the imaginary frequency $h\nu^*_1 \approx i\cdot 110$meV ($i\cdot 26.6$ THz). This indicates that the maximum of the energy in the short jump is a simple saddle point of the potential relief.

Calculations of the PhDOS for W super cell with one H atom in octapore, which is the top of the potential barrier for a long jump (0→5→6 e.g.), demonstrates one mode for the H atoms with the real frequency $h\nu^{**}_3 \approx 290$ meV (70.1 THz) and two modes for the H atom with about equal imaginary frequencies $h\nu^{**}_1 \approx h\nu^{**}_2 \approx i\cdot 100$ meV ($i\cdot 24.2$ THz). Thus, the octapore (position 5) is a saddle point of the second order in the potential relief. It is surrounded by four equivalent tetrahedral sites and four valleys, and therefore has only one real frequency of H atom in this position, in contrast to the top of the barrier for a short jump.

Physically, this is due to the feature of the potential barrier vertex (points 5) for the long jump (0-6). This top is the saddle of the second order, as is surrounded by 4 equivalent positions (1,2,3,4), i.e. it has 4 valleys of the potential relief.

3. Diffusion coefficient

The diffusion coefficient D is defined as $D = D_0 \exp(-\Delta H_m / kT)$; $D_0 = \nu \cdot \lambda^2 \cdot \exp(-\Delta S_m / kT)$, where ΔH_m is the activation enthalpy for diffusion, ΔS_m is the activation entropy for diffusion. We will treat factor D_0 similar to [13,14]. In the theory of absolute rates of reactions [13], factor D_0 includes parameter $\Gamma_j(T)$, which is the ratio of the number of allowed states on the top of the reaction barrier to the number of possible states in the equilibrium state. This relationship can be represented through the number of states for a quantum oscillator. An expression for the frequency of jumps through a barrier ε was obtained in [14] basing on [13] using the approximation of a quantum harmonic oscillator with a spectrum, for which all states are allowed with $h\nu(n+1/2)$, (n=0...∞).

Our calculations for the ground state of a quantum particle (n=0) give that $0.5\cdot\sum_{i=1}^{3} h\nu_i = 267 meV$, i.e. the energy level of H atom in the potential well corresponding to the ground state is of about the same value as the height of the barrier. Similarly, zero vibrations change the energy state of the atom in saddle points on the top of the barriers, but to a smaller extent, since the number of real frequencies is smaller. In our case, $1/2\cdot\Sigma^2_{i=1}(h\nu^*_i) \approx 225 meV$ for short jumps (0→1) and $1/2\cdot\Sigma^1_{i=1}(h\nu^{**}_i) \approx 145 meV$ for long jumps (0→6). The correction to the zero oscillations of the H atom leads to effective decrease of the activation barriers by $\Delta_1(0\to1) = -0.042 eV$ and $\Delta_2(0\to6) = -0.122 eV$. As a result, one may expect that the effective activation barrier for the long jump 0-6 is reduced approximately down to $\varepsilon_{2eff} \approx \varepsilon_2 + \Delta_2 \approx 0.25 eV$ and that for the short jump 0-1 remains about the same $\varepsilon_{1eff} \approx \varepsilon_1 + \Delta_1 \approx 0.19 eV$.

In contrast to [14], we have taken into account only the ground state of the quantum oscillator (*n=0*) and the contribution of the first excited state (*n=1*), since the higher levels of vibrations of the oscillator will exceed the height of the barrier. Taking into account only the two states for the short jump (i.e., the mode *n= (000)* and modes *(100), (010), (001)*), we have:

$$\Gamma_{(0-1)TST}(T) = \frac{kT}{h}\cdot\frac{e^{-\frac{(h\nu^*_1 + h\nu^*_2)}{2kT}}}{e^{-\frac{(h\nu_1 + h\nu_2 + h\nu_3)}{2kT}}}\cdot\frac{(1+e^{-h\nu^*_1/kT})\cdot(1+e^{-h\nu^*_2/kT})}{(1+e^{-h\nu_1/kT})\cdot(1+e^{-h\nu_2/kT})\cdot(1+e^{-h\nu_3/kT})} \sim e^{-\frac{\Delta_1}{2kT}} \quad (1)$$

A similar expression is obtained for the long jump, but only with one frequency:

$$\Gamma_{(1-6)TST}(T) = \frac{kT}{h}\cdot\frac{e^{-\frac{(h\nu^{**}_1)}{2kT}}}{e^{-\frac{(h\nu_1 + h\nu_2 + h\nu_3)}{2kT}}}\cdot\frac{(1+e^{-h\nu^{**}_1/kT})}{(1+e^{-h\nu_1/kT})\cdot(1+e^{-h\nu_2/kT})\cdot(1+e^{-h\nu_3/kT})} \sim e^{-\frac{\Delta_2}{2kT}} \quad (2)$$

The diffusion coefficients for two types of jumps D were calculated using the standard expression

$$D_j(T) = \frac{1}{6}\lambda_i^2 \cdot n_j \cdot \Gamma_j(T) \cdot \exp(-\varepsilon_j / kT) \quad (3)$$

where λ_j is the length of the jump (λ_1 =1.13 Å, λ_2= 1.6 Å); n_j is the number of nearest sites to jump. Activation energies were taken from DFT calculations $\varepsilon_1 \approx 0.195$ eV and $\varepsilon_2 \approx 0.37$ eV. Figure 4 shows the temperature dependencies of the diffusion coefficients calculated for the short and long jumps along with experimental data obtained in [1-3]. Results of calculations given in [11] are also shown for comparison. One can see that long jumps give a good agreement with experimental data obtained in [1] at high temperatures. Short jumps with smaller activation energies overestimate diffusivity. One must also remind that directions of vibrations discussed above are also in favor of long jumps.

The Arrhenius plots 1 and 2 in figure 4 are not exactly straight lines, nevertheless one may introduce a simple exponential approximation for the apparent diffusion coefficient $D(T)$ = $2.3\times10^{-7}\exp(-0.29/kT)$ $m^{-2}s^{-1}$ shown by dashed line 4 in figure 4. The apparent activation energy 0.29 eV is less than $\varepsilon_2 = 0.37$ eV obtained in DFT. The reason is that parameter $\Gamma(T)$ depends on temperature, and vibrations reduce the effective activation energy roughly down to 0.25 eV for long jumps as it was discussed above. One must also mention that Fraunfelder's experimental data can be divided in two groups at 1250-1400 K and at 1400-2000 K with different inclinations, while the energy 0.39 eV is used as an average value for the whole range 1250-2000 K. Our apparent activation energy 0.29 eV describes the high temperature wing well. The low temperature wing can be affected by defects as it was discussed in [11,15].

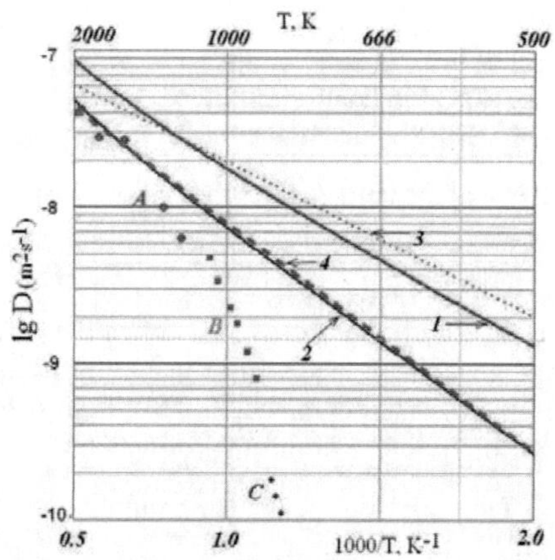

Figure 4. Temperature dependences of the diffusion coefficient D for the short (1) and long (2) jumps. Dotted line 3 – calculations [11]. Dashed line 4 is a simple exponential approximation. Experimental data: • - from [1]; ■ - from [2]; ♦ - from [3].

4. Conclusions

Potential energy of the W-H system with H atom jumping between two tetrahedral positions was calculated by DFT in the temperature range of 500-2000K for two types of jumps: a short jump between two nearest tetrahedral positions, and a longer jump through an intermediate octahedral position. The difference of energies in equilibrium tetrahedral position and in saddle points was $\varepsilon_1 \approx 0.197$ eV for the short jump and $\varepsilon_2 \approx 0.37$ eV for the long jump. Zero vibrations of atoms enhance the energy of the atom and reduce the effective activation energy down to roughly $\varepsilon_{1\text{eff}} \sim 0.19$ eV and $\varepsilon_{2\text{eff}} \sim 0.25$ eV, respectively. The diffusion coefficient calculated for long jumps is in a good agreement with experimental data of Fraunfelder in the range of 1400-2000 K and can be approximated by apparent diffusion coefficient $D(T) = 2.3\times10^{-7}\exp(-0.29/kT)$ $m^{-2}s^{-1}$. The experimental points in this temperature range can be described by smaller apparent activation energy than the activation energy 0.39 eV obtained in a wider temperature range. Short jumps overestimate diffusivity. Analyses of vibration directions is also in favor of long jumps, as vibrations of H atom take place in the directions of these jumps, while there are no vibrations in the directions of short jumps.

Acknowledgements
This work was supported by National Research Nuclear University MEPhI in the framework of the Russian Academic Excellence Project (contract No. 02.a03.21.0005, 27.08.2013).

References

Journal of Physics: Conference Series **748** (2016) 012010 doi:10.1088/1742-6596/748/1/012010

[1] Frauenfelder R, 1969 *J. Vac. Sci. Tech.* **6** 388
[2] Zakharov A P, Sharapov V M, Evko E I 1973 *Soviet. Mater. Sci.* **9** 149
[3] Benamati G, Serra E, Wu C H 2000 *J. Nucl. Mat.* **283-287** 1033
[4] Moore G E, Unterwald F C 1964 *J. Chem. Phys.* **40** 2639
[5] Ryabchikov L N 1964 *Ukr. Fiz. Zh.* **9** 293
[6] Ikeda K, Otsuka T, Tanabe T 2011 *J. Nucl. Mater.* **415** S684
[7] Otsuka T, Hoshihira T, Tanabe T 2009 *Phys. Scr.* **T138** 014052
[8] Hoshihira T, Otsuka T, Wakabayashi R, Tanabe T 2011 *J.Nucl. Mater.* **417** 559
[9] Liu Y L, Shi W 2 A.013 *Fusion Eng. Des.* **88** A 368
[10] Heinola K, Ahigren T 2010 *J. Appl.Phys.* **107** 113531
[11] Xiang-Shan Kong, Sheng Wang, Xuebang Wu et al 2015 *Acta Materialia* **84** 426
[12] Degtyarenko N, Pisarev A 2015 *Physics Procedia* **71** 30
[13] Glasstone S, Eyring H, Laidler K J 1941 *The theory of rate processes* (McGraw-Hill)
[14] Fernandez N, Ferro Y and Kato D 2015 Acta Materialia **94** 307 http://dx.doi.org/10.1016/j.actamat.2015.04.052
[15] Oda T, Zhu D, Watanabe Y 2015 *J. Nucl. Mater.* **467** 439

Effect of irradiation by argon ions on hydrogen transport through the surface oxide layer of zirconium

A E Evsin[1], L B Begrambekov[1], A I Gumarov[2], N F Kashapov[2], A G Luchkin[2], I R Vakhitov[2], I V Yanilkin[2] and L R Tagirov[2]

[1]National Research Nuclear University MEPhI (Moscow Engineering Physics Institute), 31 Kashirskoe highway, Moscow 115409, Russia
[2]Kazan Federal University, 18 Kremlevskaya st., Kazan 420008, Russia

Corresponding author e-mail: evsin@plasma.mephi.ru

Abstract. Effect of zirconium irradiation by 1 keV Ar^+ ions on hydrogen transport through the surface oxide layer is studied. It is shown that deuterium trapping under subsequent irradiation of the Ar-treated sample by deuterium atoms of thermal energies in D_2 + 30at.% O_2 gas mixture is 2 times less than trapping in the untreated sample. Besides, irradiation of the untreated sample by D-atoms provokes desorption of ≈25% of hydrogen contained therein, whereas hydrogen desorption from the ion-treated zirconium surface does not occur. It is proposed that oxygen depletion of the surface oxide layer, caused by ion bombardment, is a reason of mitigation of the hydrogen transport through this layer in both directions.

1. Introduction

Zirconium components of the active core of light-water reactors absorb hydrogen due to their interaction with water coolant and because of irradiation by hydrogen atoms and ions produced at radiolysis of water [1, 2]. Ion and atomic irradiation intensifies hydrogen transport through the surface oxide layer of zirconium in both directions causing desorption of part of the hydrogen atoms from zirconium, but, at the same time, accelerating hydrogen absorption [3-5]. Accumulation of hydrogen in zirconium components leads to embrittlement and subsequent degradation of zirconium. These processes limit reactor lifetime and fuel burn up. Thereby, methods of surface passivation against hydrogen penetration into zirconium are actively developed. Some authors [6, 7] observed deceleration of the ion-irradiated zirconium hydrogenation during exposure to overheated steam and ambient water. However, a mechanism of mitigation of the hydrogen penetration is not enough studied. In this work, the effect of Ar^+ irradiation of zirconium on hydrogen transport through the surface oxide layer under irradiation by deuterium atoms of thermal energies was investigated employing thermal desorption spectroscopy (TDS) and X-ray photoelectron spectroscopy (XPS).

2. Experimental

The investigated samples were the tube sections with a size of ≈7×7×1 mm^3 made of zirconium alloy E110 (Zr–1%Nb) preliminarily rinsed in ultrasonic ethanol bath. Some of the samples were further irradiated by ions of argon plasma. After that, both ion-treated and untreated samples were irradiated by deuterium atoms of thermal energies in deuterium-oxygen gas mixture. All the experiments and TDS analysis were performed in MIKMA device [8].

The argon plasma irradiation parameters were as follows: the ions with the energy of 1 keV were delivered at the flux of 1.1×10^{20} $m^{-2}s^{-1}$ to the fluence of 1.3×10^{23} m^{-2}. Residual gas pressure in the plasma chamber did not exceed 1.3×10^{-3} Pa, the main component of the residual gas was water molecules (>90%). The argon pressure before ignition of the discharge was set at 1.3×10^{-1} Pa. The sample temperature during irradiation was maintained at 550 K.

Irradiation by deuterium atoms of thermal energies was carried out by the device [9], where deuterium atoms were generated by means of heating the tungsten spiral up to ≈1800 K in gaseous deuterium. Parameters of the experiments were as follows: the operating gas – a mixture D_2+30%O_2 with the deuterium pressure of 1.2×10^{-1} Pa. Molecular flux onto the surface of the tungsten spiral, dissociation probability of deuterium molecule on the surface of tungsten at given temperature and solid angle, at which each section of the spiral "sees" the sample, were taken into account in the calculation of the atomic flux. The latter was estimated as ~10^{19} $m^{-2}s^{-1}$, and the fluence ~10^{23} m^{-2}. The sample temperature during atomic irradiation was maintained at 600 K.

The residual gas pressure in the TDS chamber during analysis did not exceed 7×10^{-6} Pa, while the sample was heated linearly at a rate of 5 K/s in the temperature range of 300-1500 K. Desorption of H_2, HD, D_2, H_2O, HDO, D_2O was registered during the analysis. Desorption of other molecules, containing hydrogen isotopes, was negligible. The amounts of hydrogen and deuterium in zirconium were obtained by integration of the TDS spectra of H-containing and D-containing molecules.

XPS depth profiling was performed in SPECS facility [10]. All the measurements were carried out at room temperature. The energy and current of etching ions were set to 3 keV and 10 μA, respectively. Zr3d and O1s photoemission spectra were studied.

3. Results and discussion

Results of the XPS measurements have shown that irradiation of zirconium by ions of argon plasma leads to the growth of contribution of oxide components in the XPS Zr3d spectra throughout the investigated depth (figure 1, a). It should be interpreted as a growth of the oxide layer. At the same time, the ratio of ZrO_2 to all zirconium oxide components ZrO_x ($x \leq 2$) in the Zr3d spectrum near the surface of the ion-treated sample decreases (figure 1, b).This alteration indicates the appearance of the oxygen defects (anion vacancies or others) in the surface layer under the ion bombardment.

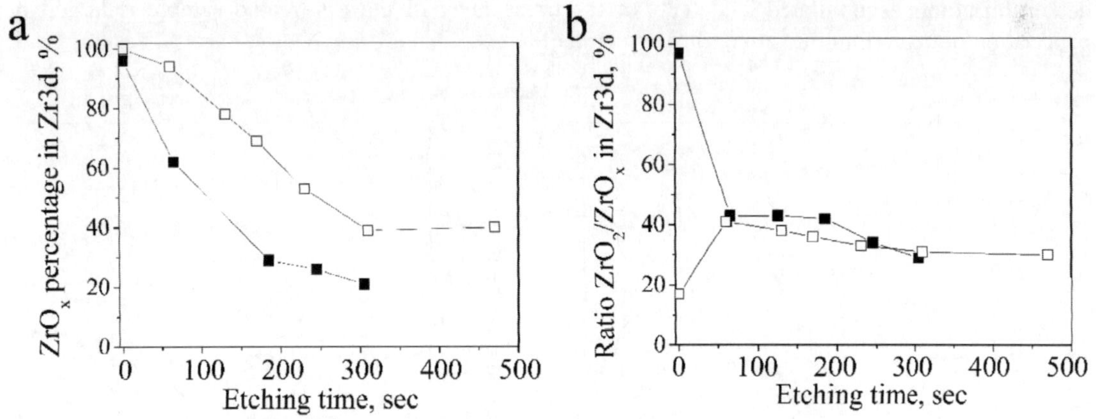

Figure 1. a) ZrO_x ($x \leq 2$) percentage in the XPS spectrum Zr3d versus etching time of the ion-treated sample (□) and the untreated one (■); b) Ratio of ZrO_2 to all zirconium oxide components ZrO_x ($x \leq 2$) in the XPS spectrum Zr3d versus etching time of the ion-treated sample (□) and the untreated one (■).

Irradiation of the both ion-treated and untreated samples by deuterium atoms of thermal energies in deuterium-oxygen gas mixture leads to growth of the oxide layer to approximately equal thickness. At the same time, percentage of oxide components in XPS spectrum Zr3d (figure 2, a) and the ratio of ZrO_2 to all zirconium oxide components ZrO_x ($x \leq 2$) (figure 2, b) near the surface of the ion-treated sample is less than that of the untreated one. Thus, it could be proposed that defects, generated by the ion bombardment of zirconium, remain in the oxide layer after subsequent irradiation by deuterium atoms of thermal energies in the gas mixture D_2+30%O_2.

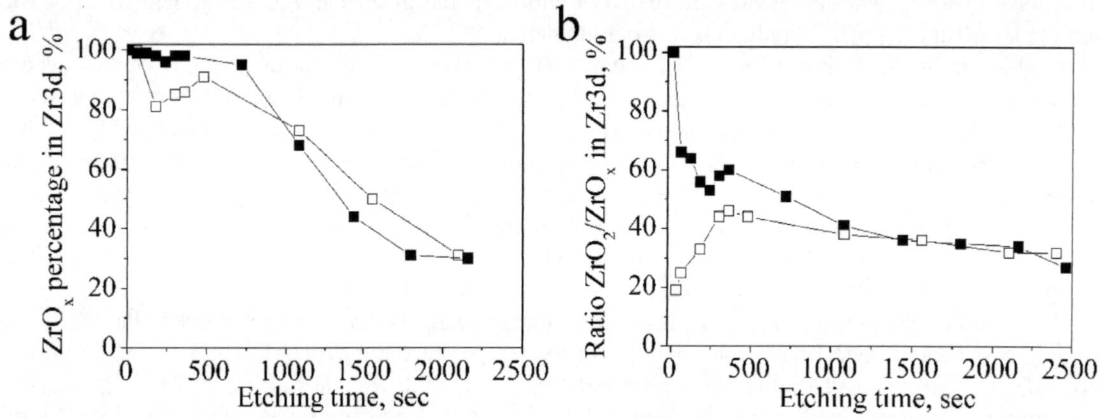

Figure 2. a) ZrO_x ($x \leq 2$) percentage in the XPS spectrum Zr3d versus etching time of the ion-treated sample (□) and the untreated one (■) after irradiation by D-atoms in D_2+30%O_2; b) Ratio of ZrO_2 to all zirconium oxide components ZrO_x ($x \leq 2$) in the XPS spectrum Zr3d versus etching time of the ion-treated sample (□) and the untreated one (■) after irradiation by D-atoms in D_2+30%O_2.

Deuterium trapping under atomic irradiation of the untreated sample is equal to 3.4×10^{21} m^{-2}, whereas the ion-treated sample absorbs ≈2 times fewer deuterium atoms. Maximum at 1050 K in the TDS spectrum of deuterium-containing molecules from the ion-treated sample is higher, and the peak at 1300 K is lower than the corresponding maxima in the TDS spectrum obtained from the untreated sample (figure 3, a). According to interpretation given in our previous works (e.g., [5]), the peak at 1300 K testifies to desorption of deuterium retained in the bulk of zirconium, whereas the maximum at 1050 K corresponds to desorption from trapping centers in the oxide layer. Thus, the amount of deuterium atoms accumulated in defects of the oxide layer of the ion-treated sample is larger, but penetration of deuterium into zirconium is retarded.

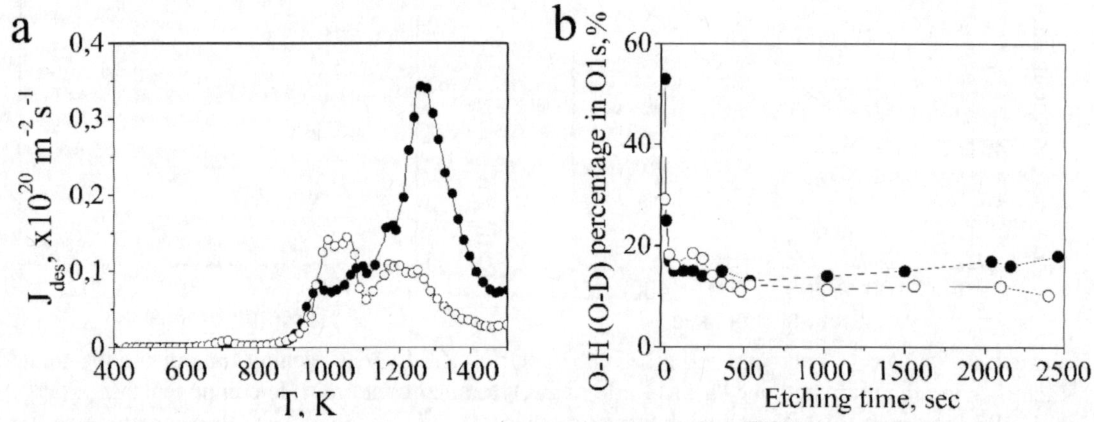

Figure 3. a) TDS spectra of deuterium-containing molecules from the ion-treated sample (○) and the untreated one (●) after irradiation by D-atoms in D_2+30%O_2; b) O-H (O-D) percentage in the XPS spectrum O1s versus etching time of the ion-treated sample (○) and the untreated one (●) after irradiation by D-atoms in D_2+30%O_2.

Irradiation of the untreated zirconium by D-atoms provokes desorption of ≈25% of hydrogen atoms contained therein. In contrast, hydrogen desorption during irradiation of the ion-treated zirconium is not observed. It should be also mentioned that percentage of hydroxyl component (that may contain either H or D atoms) in the XPS spectrum O1s in the depth of the oxide layer of the ion-treated sample after irradiation by D-atoms is less than that of the untreated sample (figure 3, b).

Based on the described results, one could assume that retardation of the hydrogen transport through zirconium oxide layer irradiated by ions of argon plasma occurs due to the presence of oxygen defects produced by ion bombardment. Hydrogen in zirconia diffuses by hopping between adjacent oxygen ions [11], so the decrease of the oxygen concentration in the oxide layer caused by the ion bombardment leads to increase of the mean distance between adjacent oxygen ions. As a result, probability of the hopping decreases and hence, the hydrogen transport decelerates.

4. Conclusion

Experimental results have shown that irradiation of zirconium by argon ions with the energy of 1 keV leads to decrease of the oxygen concentration in the surface oxide layer, apparently, due to production of oxygen defects (anion vacancies or others). Defectiveness of the surface oxide layer remains after irradiation of the ion-treated zirconium by deuterium atoms with thermal energies in the deuterium-oxygen gas mixture. Deuterium trapping in the ion-treated zirconium under irradiation by D-atoms is 2 times less than that of the untreated sample. Hydrogen desorption from the ion-treated zirconium under atomic irradiation is not observed, and the oxide layer contains fewer amount of hydroxyl groups. One could propose that oxygen depletion of the surface layer, caused by ion bombardment, is the reason of retardation of the hydrogen transport through zirconium oxide layer in both, in and out, directions.

Acknowledgments

A E Evsin and L B Begrambekov were supported by the Ministry of Education and Science of the Russian Federation (agreement contract No 14.575.21.0049 from 27.06.14) and by National Research Nuclear University MEPhI in the framework of the Russian Academic Excellence Project (contract No. 02.a03.21.0005, 27.08.2013). A I Gumarov, N F Kashapov, A G Luchkin, I R Vakhitov, I V Yanilkin and L R Tagirov were supported by the Program of Competitive Growth of Kazan Federal University. XPS measurements were carried out at the PCR Federal Center of Shared Facilities of KFU.

References

[1] IAEA-TECDOC-996 1998 *Waterside corrosion of zirconium alloys in nuclear power plants* (Vienna: International Atomic Energy Agency) 91-123
[2] Burns W G and Moore P B 1976 *Radiat. Eff. Defects Solids* **30** 233
[3] Axelsson K-O, Keck K-E and Kasemo B 1986 *Appl. Surf. Sci.* **25** 217
[4] Begrambekov L, Dvoychenkova O, Evsin A, Kaplevsky A, Sadovskiy Ya, Schitov N, Vergasov S and Yurkov D 2014 *J. Phys. Conf. Ser.* **567** 012003
[5] Evsin A E, Begrambekov L B, Gumarov A I, Kaplevsky A S, Luchkin A G, Tagirov L R and Vakhitov I R 2016 *Vacuum* **129** 183
[6] Une K, Takagi I, Sawada K, Watanabe H, Sakamoto K and Aomi M 2012 *J. Nucl. Mater.* **420** 445
[7] Chernov I P, Berezneeva E V, Beloglazova P A, Ivanova S V, Kireeva I V, Lider A M, Remnev G E, Pushilina N S and Cherdantsev Y P 2014 *Tech. Phys.* **59** 535
[8] Airapetov A, Begrambekov L, Brémond S, Douai D, Kuzmin A, Sadovskiy Ya, Shigin P and Vergasov S 2011 *J. Nucl. Mater.* **415** S1042
[9] Russian patent RU 2479167 2013
[10] Vakhitov I R, Lyadov N M, Valeev V F, Nuzhdin V I, Tagirov L R and Khaibullin R I 2014 *J. Phys. Conf. Ser.* **572** 012048

[11] Rabone J, Van Uffelen P 2015 *J. Nucl. Mater.* **459** 30

Journal of Physics: Conference Series **748** (2016) 012012 doi:10.1088/1742-6596/748/1/012012

Arc tracks on nanostructured surfaces after microbreakdowns

D Sinelnikov[1], D Bulgadaryan[1], D Hwangbo[2], S Kajita[3], D Kolodko[1], V Kurnaev[1] and N Ohno[2]

[1]National Research Nuclear University MEPhI (Moscow Engineering Physics Institute), 31 Kashirskoe highway, Moscow 115409, Russia
[2]Graduate School of Engineering, Nagoya University, Nagoya 464-0803, Japan
[3]Institute of Materials and Systems for Sustainability, Nagoya University, Nagoya 464-8603, Japan

Corresponding author e-mail: dnsinelnikov@mephi.ru

Abstract. Studying of initial steps of unipolar arc ignition process is important for reduction of probability of arcing between the plasma and the wall in thermonuclear devices. Tungsten nano-fuzz surface formed by helium plasma irradiation at high fluences and temperatures is a perfect material for arc ignition. Snowflake-like craters were detected on the fuzzy surfaces after short micro-breakdowns. Such sort of craters have not been observed before on any other metallic surfaces. These specific traces are formed due to unique properties of the fuzz structure. The nano-fuzz could be easily melted and vaporized by micro-breakdown current, due to its porosity and bad thermal conductivity, and formation of low conducting metallic vapour under the cathode spot causes discharge movement to the nearest place. Thus, even low current arc can easily move and leave traces, which could be easily observed by a secondary electron microscope.

1. Introduction

Unipolar arcing is a negative phenomenon taking place between the plasma and the first wall in fusion devices with magnetic confinement. Moreover, arcing becomes more frequent with increasing the surrounding plasma density. It is well known that arcing probability is strongly dependent on properties of plasma facing materials [1]. Thin (10-30 nm) tungsten nanostructures called fuzz can be formed on tungsten surfaces after helium irradiation with fluence of ~10^{25} m^{-2} when the surface temperatures was in the range of 1000-2000 K [1]. It was shown [2] that fuzzy tungsten surfaces can demonstrate enhanced frequency of arcing in linear simulator machines, but there are only few experimental works that investigated the mechanisms of such behavior. Thus, investigation of the pre-breakdown phenomena is an important step for mitigating unipolar arcing. In this study, pre-breakdown phenomena on the fuzzy tungsten samples are investigated using vacuum diode measurements.

2. Experimental setup

Vacuum diode designed for current-voltage characteristic (CVC) measurements from fuzzy surfaces is shown in figure 1(a). This device allows comparing CVCs of different fuzzy structures grown on substrates of equal thickness without changing of the vacuum gap. The gap distance can be varied from 0.3 mm, which allows working with the electric field strength up to 40 kV/mm using voltage source up to 12.5 kV and maximum current of 2 mA. The spherical shape of the anode is for solving the problem of parallel positioning of the electrodes that is very important for small distances. The vacuum and electric schemes of the setup are shown in figure 1 (b). The operating vacuum pressure was in the range of 3-5 10^{-7} Torr for restricting secondary processes in the vacuum gap. Electrical scheme consisted of the negative high voltage source 7, nanoammeter 8 and ballast resistor 6 for limiting current at the 300 μA level in case of a breakdown, as shown in figure 1(b).

a)

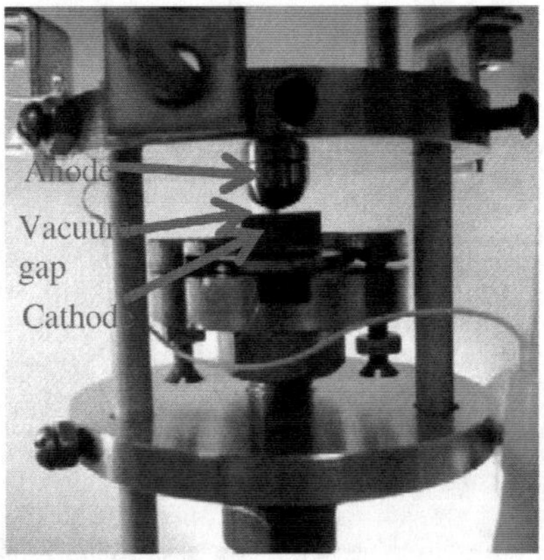

b)

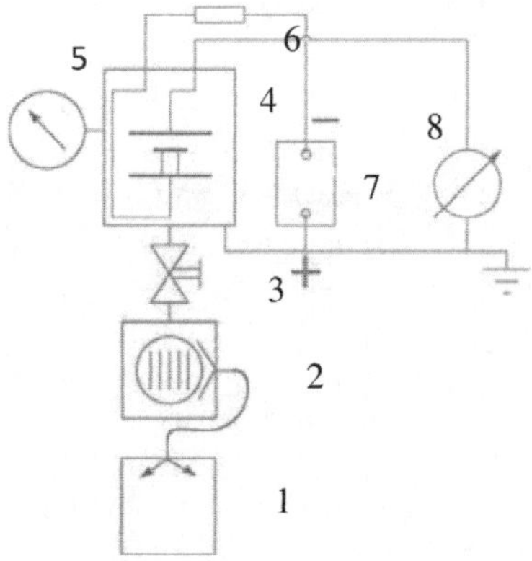

Figure 1. A photo of the vacuum diode for pre-breakdown current measurements (a) and the scheme of setup (b): 1 – rotary pump, 2 – turbomolecular pump, 3 – gate, 4 – vacuum chamber, 5 – pressure gauge, 6 – ballast 25 MOhm resistor, 7 – high voltage source, 8 – nanoammeter.

3. Results and discussion

Fuzzy samples were prepared in the NAGDIS-II device [3] with high density helium plasma, the typical density and temperature of which are, respectively, $10^{19}m^{-3}$ and 5 eV. Specimens were installed to the plasma and electrically biased to −100 V for reaching total fluence ~10^{25} m^{-2}. The surface temperature was raised by the bombardment of helium ions up to 1000 K. A CVC was measured in the vacuum diode just after preparation in the NAGDIS-II. Only small nanoampere currents of 25-35 kV/mm were detected and are shown on the Fowler-Nordheim plot in figure 2 (black dots).

After increasing of the field strength up to 40 kV/mm, a microbreakdown has occurred, and after this event, CVC changed: significantly higher currents up to several tens of microamperes were emitted at much lower field strength (gray dots in figure 2). The currents have not changed even after an exposure in atmosphere and the reason seems to be only modification of the surface relief during the microbreakdown.

Local field amplification factors were calculated using CVCs for two cases: before and after the breakdown. According to the equation (1) for relation between the emission current I and the electric field strength E, the local field enhancement factor β can be calculated from the slope [4].

$$\lg\left(\frac{I}{E^2}\right)=-\frac{2.818\cdot10^9\cdot\varphi^{3/2}}{\beta}\cdot\frac{1}{E}+\frac{4.406}{\varphi^{1/2}}+\lg\left(\frac{S\cdot\beta^2}{\varphi}\right)-5.854 \quad (1)$$

In this equation φ=4.5eV is the work function for tungsten, S is effective emission area. Before the breakdown this factor was 110 and taking into account that β can be estimated roughly for a tip on the plane as h/r (h - height of the tip, r - radius of the tip), such value can be rather possible for some protrude nanostructures. But in the case of "after the breakdown", β was about 1300, and this value seems impossibly high for surface protrusion. These currents could not be explained by the classical field emission theory and could be due to thermofield emission from heated protrusions.

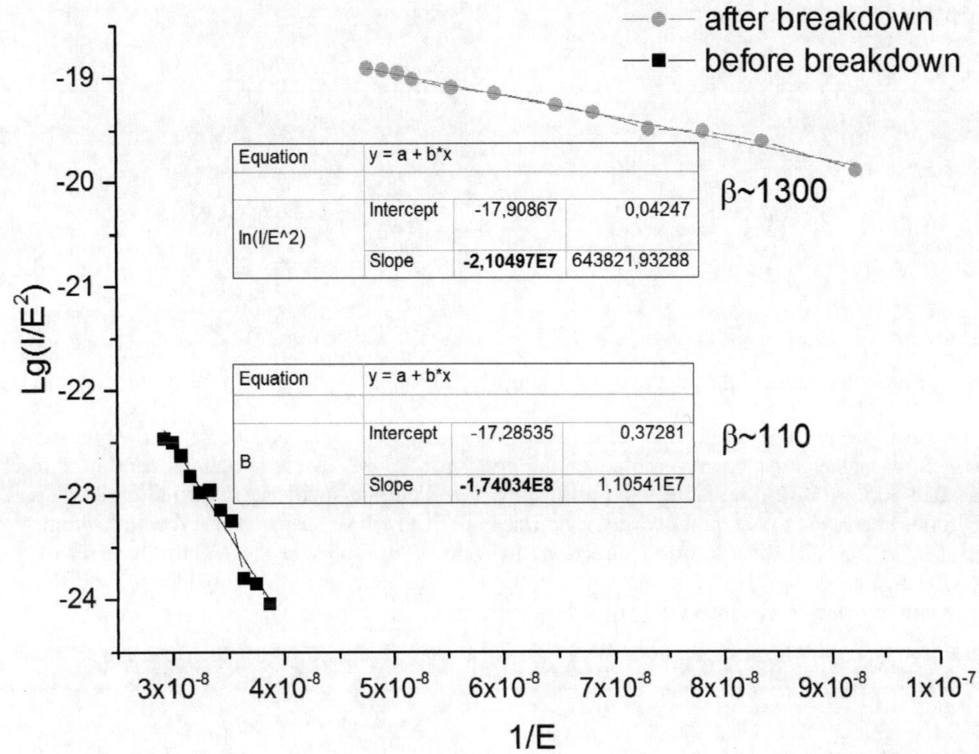

Figure 2. Current-voltage characteristic of the tungsten nanostructured sample before and after a microbreakdown.

After field emission measurement, the emission zone of the fuzz sample was observed by a scanning electron microscope (SEM). Some parameters of the breakdown are shown in table 1.

Table 1. Parameters of the microbreakdowns on tungsten and molybdenum.

Material of fuzzy structure	W	Mo
Break voltage	10 kV	12.5 kV
Current after break	163 μA	236 μA
Number of large "snowflakes"	12	55
Diameter of "snowflake"	50 μm	50 μm

The places of microbreakdown are shown in figure 3 for tungsten (a) and molybdenum (b) fuzz. The area modified by craters is about 1.5 mm^2, which is rather close to the calculated one for the used vacuum diode with the spherical anode. In this calculation the area of the cathode, which was estimated near which electric field strength, is not lower than 90% from maximum value. The breakdown region consists of two separated parts shown in figure 3, the main black crater with a lot of black dust around it and a lot of "snowflake"-like craters in some distance from the main one. The reason of separation can be due to local field reduction near the main crater during the breakdown process. Probably, the main crater is formed earlier than "snowflakes", because it is surrounded by them. The main crater looks like after explosion due to a lot of small black dust near it and this explosion can be the initiator of the breakdown.

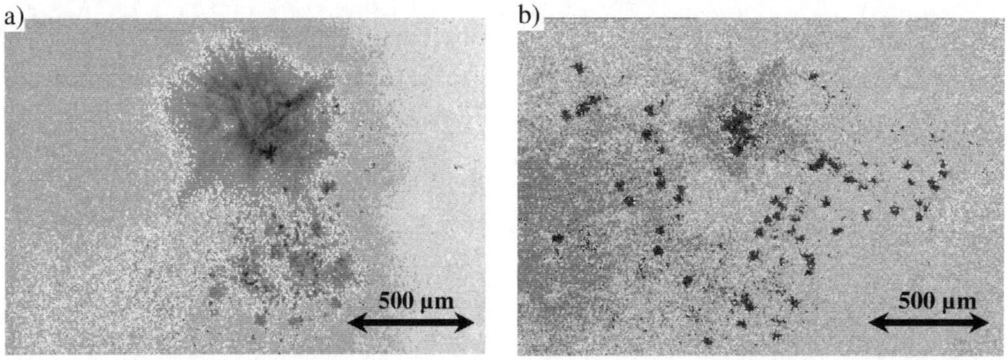

Figure 3. Places of the breakdown for tungsten (a) and molybdenum (b) fuzzy samples.

However, the presence of the main crater is not necessary for the cathode area after the breakdown. In figure 4 another breakdown place contains only "snowflakes". These structures consist of black separated dots with ~2 µm diameter – presumably they were arc spot traces, but these black points very often merge into continuous tracks. The tracks of "snowflake" usually have central symmetry. This means that arcs diverge in all directions in spite of surface scratches. The diameters of the snowflakes are usually about 50 µm, which is the same for tungsten and molybdenum and doesn't depend on the nanostructure material properties.

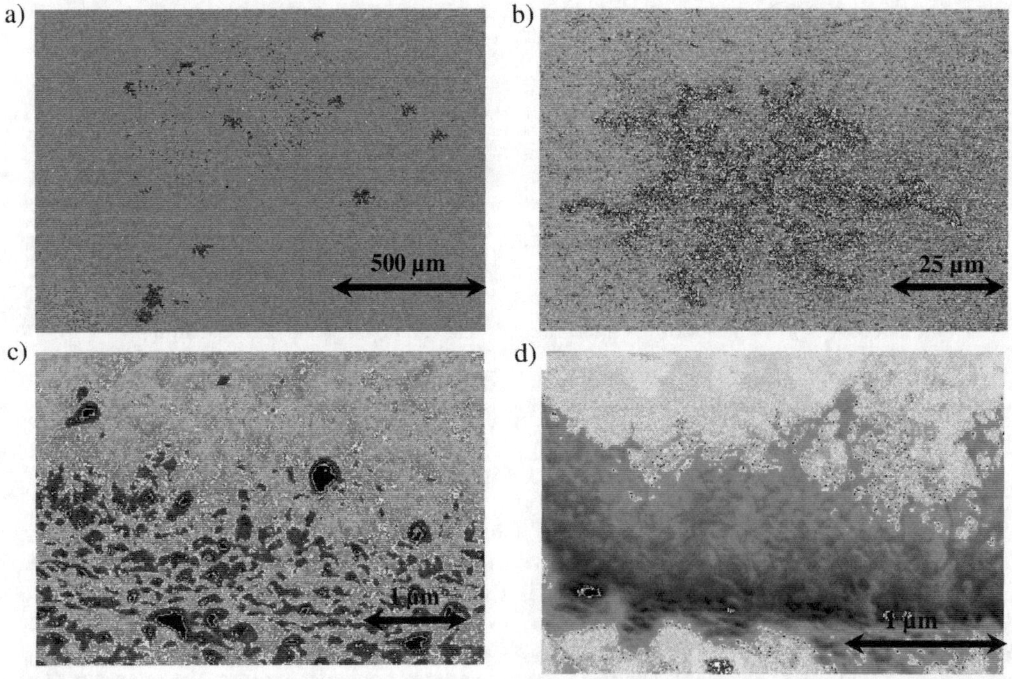

Figure 4. Photos of the "snowflakes" on the tungsten fuzz surface after a breakdown: a) and b) general view of the snowflakes, c) and d) the sliding angle view on the "snowflake" crater bottom.

The arc current is usually rather high compared with the current limit by the ballast resistor, so the capacity of the vacuum diode could be the main source of the current. The last one can be estimated about 10^{-12} F.

When the arc spot is hot enough for evaporating of liquid material of the cathode, it is usually profitable for the arc to move on to the nearest place, because of the bad conductivity of the metal vapor and lower resistant of a new cold place at all [5]. Thus, the necessary energy for evaporation of the tracks of the "snowflakes" can be comparable with the total capacitor energy. Then, the crater depth can be estimated by the order of value in equation (2). For this estimation, the density of fuzzy structure was taken 0.05 of tungsten density from [6]. Other referential parameters were taken from [7] and are shown in table 2.

$$H \cdot S \cdot 0.05\rho \cdot (c \cdot (T_{\text{melt}}\text{-}300\text{K}) + H_{\text{melt}} + c \cdot (T_{\text{boil}} - T_{\text{melt}}) + H_{\text{boil}}) = CU^2/2 \quad (2)$$

Table 2. Number of craters and material properties of nanostructure.

Material of the fuzzy sample	Total "snowflakes" area S, μm^2	Density ρ, kg/m^3	Heat capacity c, kJ/kg·K	Melting point T_{melt}, K	Boiling point, T_{boil}, K	Heat of fusion H_{melt}, kJ/kg	Heat of vaporization H_{boil}, kJ/kg	Break-down voltage U, kV
Tungsten	10^4	19250	0.13	3700	6200	190	4200	10
Molybdenum	10^5	10280	0.25	2900	4900	390	6200	12.5

Calculation of the craters depth H by the formula (2) gives ~1 μm for tungsten fuzz and ~0.2 μm for molybdenum, which is close to the depth observed by SEM. It is interesting to note that in spite of the much higher melting point of tungsten, the amounts of energy necessary for evaporation mass unit are very close to each other due to the higher heat capacity, heat of fusion and vaporization.

For investigation of dependence between the total area of craters and the capacitive energy of the vacuum diode, a microbreakdown for 5 kV was initiated and only 4 small "snowflakes" were found on the surface after the current increase from 5 to 21 μA. The total area of the craters was ~$2\cdot10^3$ μm, which was 5 times less than that for 10 kV breakdown and well corresponds to the capacitive energy of vacuum diode with 5 kV voltage.

4. Conclusion

It was shown that surfaces covered by the fuzzy nanostructure can indicate traces of the breakdown and pre-breakdown phenomena that is very important for investigation of arc initiation. Such effect can be explained by ease of fuzz melting due to well thermal insulation of each nanostructure. Arcs are moving rather randomly on the surface that is well illustrated in "snowflake"-like traces. It was established that formation of the "snowflake" craters significantly increases intensity of the pre-breakdown currents. The energy source for the breakdown was a capacity of the vacuum diode: a correlation was found between the diode energy under fixed voltage and the number and the size of the craters. "Snowflake" craters formation accompanied by tungsten dust erosion is the negative outcome for a thermonuclear machine.

Acknowledgements
This work was supported by National Research Nuclear University MEPhI in the framework of the Russian Academic Excellence Project (contract No. 02.a03.21.0005, 27.08.2013).

References

[1] Kajita S et al 2009 *Nucl. Fusion* **49** 095005
[2] Kajita S et al 2009 *Nucl. Fusion* **49** 032002
[3] Takamura S et al 2006 *Plasma Fusion Res.* **1** 51

[4] Mesyats G 2000 *Cathode Phenomena in a Vacuum Discharge: The Breakdown, the Spark and the Arc* (Moscow: Nauka)
[5] Anders A 2008 *Cathodic Arcs: From Fractal Spots to Energetic Condensation* (New York: Springer)
[6] Hwangbo D et al 2014 *Results in Physics* **4** 33
[7] Hammond C R 2004 *The Elements*, in Handbook of Chemistry and Physics (81st ed.) (CRC press)

Journal of Physics: Conference Series **748** (2016) 012013 doi:10.1088/1742-6596/748/1/012013

Magnetron deposited TiN coatings for protection of Al-Cu-Ag-Mg-Mn alloy

Tatiana V Stepanova[1]*, Andrey V Kaziev[1], Mikhail V Atamanov[2], Alexander V Tumarkin[1], Svetlana A Dolzhikova[3], Nelly Ph Izmailova[3], Maxim M Kharkov[1], Maria M Berdnikova[1], Dmitry V Mozgrin[1] and Alexander A Pisarev[1]

[1]National Research Nuclear University MEPhI (Moscow Engineering Physics Institute), 31 Kashirskoe highway, Moscow 115409, Russia
[2]Inacotec JSC, 4b5 Bardina st., Moscow 119334, Russia
[3]Ufa Engine Industrial Association, 2 Ferina st., Ufa 450039, Russia

Corresponding author e-mail: shtanya@list.ru

Abstract. TiN coatings were deposited on a new Al super-alloy by magnetron sputtering in argon/nitrogen environment. The deposited layer structure, microhardness, adhesion, corrosion resistance, and fatigue life were investigated and tests demonstrated improved performance of the alloy.

1. Introduction

Aluminum alloys are widely used in the aircraft industry, and one of the specific applications is for compressor blades. This engine part is subjected to sand wear, atmospheric and water corrosion, and fatigue. To improve the performance of aluminum alloys in this application, three approaches are used: development of new materials, thermal treatment, and surface modification.

Surface modification is widely used in processing of steels, and plasma treatment has become a competitive technology from the point of view of product quality, technological flexibility, cost efficiency, and environmental aspects. Plasma treatment of aluminum alloys has no wide industrial application yet, though remarkable amount of papers are focused on their plasma treatment. The traditional way of improving characteristics of aluminum alloys is physical vapor deposition (PVD), particularly by magnetron sputtering, and Ti nitride demonstrates increase of hardness, corrosion resistance and wear resistance [1-3]. Presently, TiN films are successfully used for coating of the compressor blades of the jet turbine engines to inhibit erosion and enhance strength characteristics [4]. However, using solely TiN films generally does not guarantee the required quality of protection against erosion by micro particles, therefore coatings with improved anti-erosion properties are necessary [3, 5]. Erosion by micro particles can also accelerate corrosion of blades in saliferous water vapor, and in turn, corrosion of the aluminum substrate accelerates erosion of the protecting films by particles.

The temperature of aluminum alloys during deposition must be low (below 150–200°C). Therefore, TiN coatings deposited on aluminum usually have a columnar microstructure that is typical for low-temperature film deposition. This kind of the structure incorporates voids, pores, small holes and

* To whom any correspondence should be addressed.

through channels [6] that may reduce the corrosion resistance [7], coating hardness and erosion resistance. Another challenge associated with film deposition is degradation of the fatigue properties. So far, the latter issue has remained unaddressed.

NRNU MEPhI jointly with USATU and Inacotec JSC have recently devised a technique of coating for improving the wear- and corrosion-resistance of a custom aluminum super-alloy AA2139 (Al-Cu-Ag-Mg-Mn) that does not degenerate its fatigue properties. In this paper we present the results of examining the microhardness, corrosion resistance, adhesive, and fatigue properties of the Al alloy samples coated with TiN films.

2. Experimental setup

TiN coatings were deposited in a magnetron sputtering system at Inacotec JSC. A scheme and a photo of the apparatus are shown in figure 1.

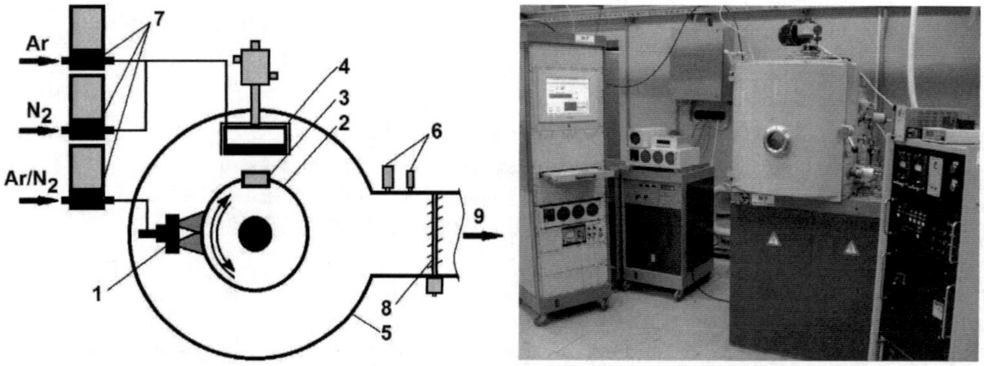

Figure 1. Experimental apparatus: *1*—ion source, *2*—sample carousel, *3*—sample holder, *4*—DC magnetron (optionally unbalanced), *5*—vacuum chamber, *6*—pressure gauges, *7*—mass-flow controllers, *8*—shutter, *9*—vacuum pumping.

The samples were attached to a sample carousel 2 in a vacuum chamber 5 at a distance of 80 mm in front of the magnetron cathode. The vacuum chamber was evacuated with a cryopump to a base pressure of 1×10^{-4} Pa. A gas-fed cold-cathode ion source 1 at a distance of 100 mm (in the processing region) had the beam cross-section area of 60×500 mm. The power supply power was 5 kW, with the maximal voltage output of 3 kV.

Prior to deposition, the samples were sputter-cleaned by 2.2–2.4 keV Ar ions (anode current 0.35–0.4 A) for 5 min at Ar pressure $p_{Ar} = 0.09$ Pa. TiN coatings were produced by reactive sputtering of Ti target in a magnetron discharge with the discharge voltage ~ 400 V (cathode current ~ 10 A) at a total Ar+N_2 pressure $p = 0.12$ Pa with the composition Ar:N_2 = 5:1. Sputter magnetron 4 operated in a pulsed DC mode at the repetition rate of 30 kHz and duty factor of 50%. During the deposition, the samples were biased with $U_b = -60$ V. Half of the samples were treated with magnetron operated in balanced regime, and the other — in unbalanced magnetron configuration. The unbalancing was induced by an auxiliary electromagnetic coil next to the magnetron's permanent magnet pack. The deposition process lasted for 10 min. Thickness of deposited films was 1 μm. The temperature of samples was measured using type E thermocouple and did not exceed 40°C. Following the deposition, the samples were cooled down to 25°C in the Ar-filled chamber at $p_{Ar} = 5\times10^{4}$ Pa.

The microstructure of samples was examined using a scanning electron microscope (SEM) Tescan Vega 3 SBH equipped with an energy-dispersive X-ray spectrometer (EDS) Oxford Instruments X-Act.

Surface hardness was measured by Future-Tech Corp. FM-800 microhardness tester. The adhesive properties of coatings were studied using Anton Paar Revetest scratch tester. Fatigue tests were carried out in an electrodynamic shaker.

The corrosion resistance of samples was investigated in express tests in 10 wt% KOH solution (pH = 15). Aluminum intensively reacts with alkali solution, so hydrogen gas bubbles appear immediately after the beginning of the tests. The sample was put on the table of an optical stereo microscope, several drops of the alkali solution were put on the sample surface, and reaction was video recorded.

3. Experimental results

3.1. Microstructure

Microstructure of the cross-sections of films prepared in balanced and unbalanced magnetron configurations, observed in SEM, is shown in figure 2. One can see that the thickness of the film is approximately the same in the two cases. There were no visible cracks and pinholes in the film.

EDS analysis showed that the coating was composed of Ti and N without impurities in the coatings.

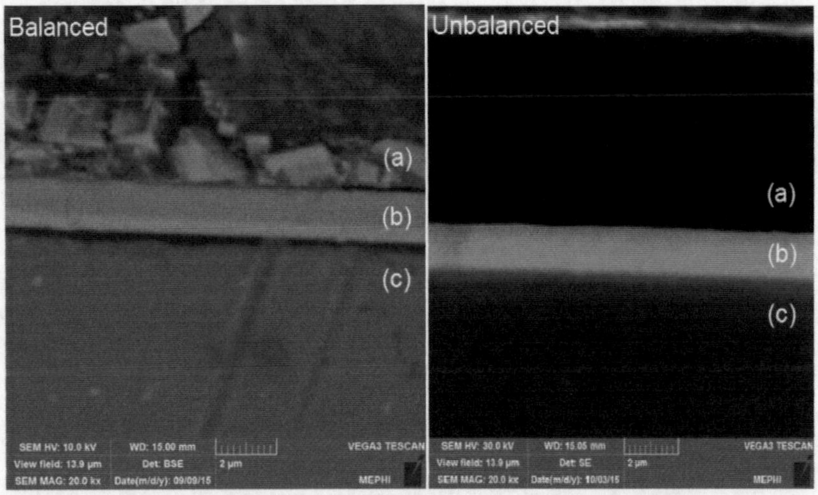

Figure 2. Cross-sections SEM images of Al alloy samples with 1-μm TiN films deposited in balanced (left) and unbalanced (right) magnetron regimes: (a) mounting, (b) TiN film, (c) Al alloy substrate.

3.2. Microhardness measurements and scratch tests

Microhardness was measured using Vickers indenter with loads normal to the surface in the range of 10–500 gf. The results of microhardness measurements as well as the SEM images of the indentation spots are given in figure 3.

One can notice that there are no cracks outside the dent, even at the maximal applied load. This indicates good adhesion of the film to the substrate.

Scratch tests were performed under progressive load, linearly increasing from 0.5 to 50 N at a rate of 49.5 N/min at the length of 5 mm with the indenter velocity of 5 mm/min. Rockwell C diamond spherical indenter tip of 200-μm radius was used. The critical loads were detected by means of an acoustic sensor and visual analyses of SEM images of scratches. Figure 4 shows an example of acoustic emission and indenter penetration during scratch test, and figure 5 shows typical images of the scratch in characteristic regions. One can see increase of the acoustic signal at the load of about 6 N, which is connected with cracking of the deposited film that can be observed in SEM images. With increase of the load, cracking increases, and the acoustic signal intensifies. One can see small cracks at the first SEM image. At the load above 15 N the signal starts to fluctuate, and SEM images demonstrate disintegration of the film: the number of cracks and the gaps between them are getting larger. Finally at the end of the scratch, the deposited film is destructed completely, and in some parts of the scratch is completely removed from the scratch bottom.

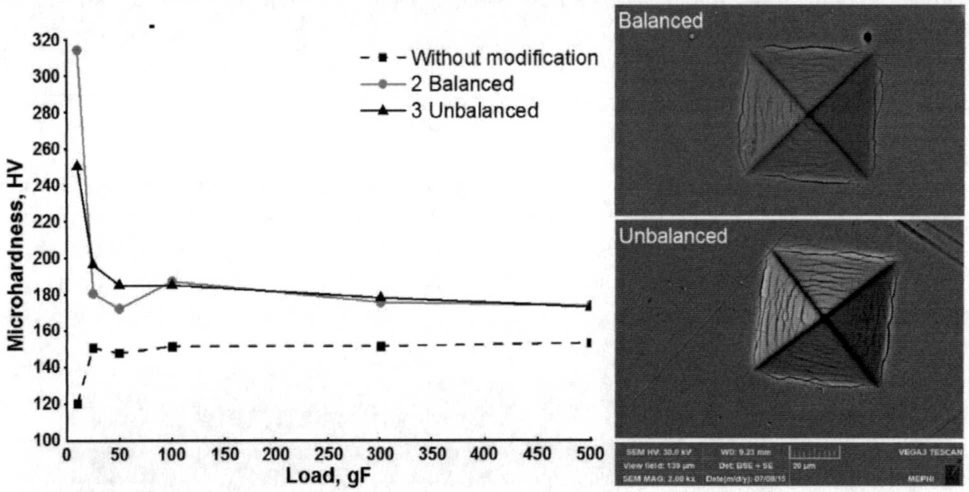

Figure 3. Surface microhardness of untreated and TiN coated samples (left); SEM images of indentations (right).

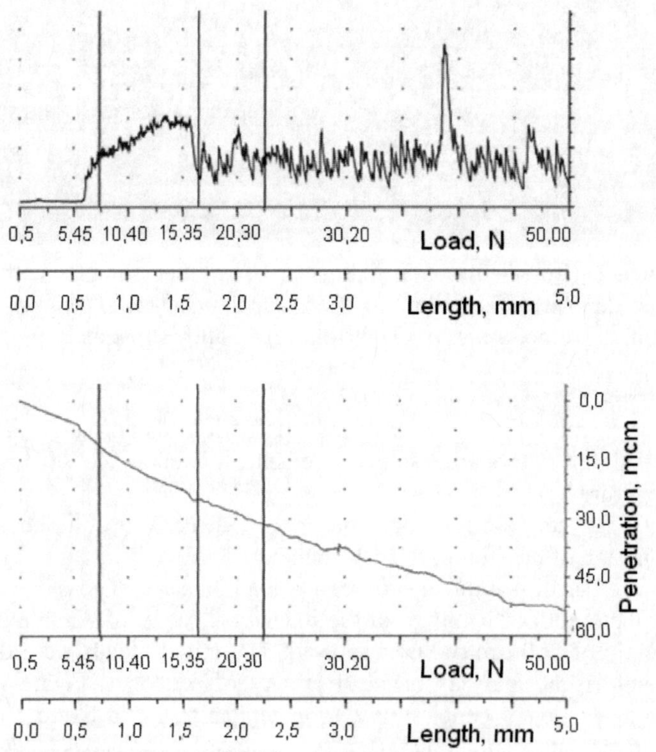

Figure 4. Acoustic signal (left) and penetration of the indenter (right) in scratch tests. The vertical lines give characteristic lengths of beginning of cracking, disintegration, and destruction of the deposited film estimated from SEM image of the scratch.

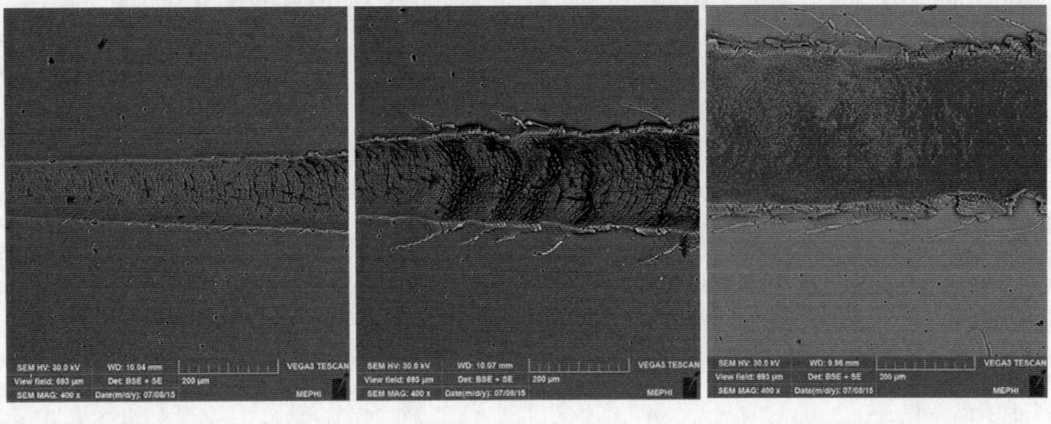

Figure 5. SEM images of the scratch in the regions of beginning of cracking (left), disintegration of the film (middle) and its destruction (right).

It is important to mention that the start of cracking leading to appearance of the acoustic signal at about 5 N happens when the indenter penetrates about 5 μm in the sample, while the thickness of the film is five times less (about 1 μm). That means that the film is rather tough under loads. It is also important that there are only small cracks outside the scratch and there is no exfoliation of the film outside the scratch even at the largest load of 50 N when the indenter penetrates 60 μm in the sample (60 times the thickness of the film).

No significant influence of the magnetron unbalancing on adhesive properties has been observed, however the performance of coatings deposited in the balanced regime was slightly better. The samples coated in the balanced magnetron regime had average characteristic critical loads of about 6 N (beginning of cracking) and about 17–18 N (intensive fracturing). The corresponding values for coatings deposited in the unbalanced regime were 5 N and 15–16 N.

3.3. Corrosion resistance

The untreated Al alloy samples exposed to 10 wt% KOH solution exhibited intensive formation of gas bubbles in less than 1 s after the beginning of the test. After 10 min, severe erosion of the surface was observed. TiN coatings significantly inhibited the corrosion of the Al alloy. Formation of gas bubbles on the surfaces coated with 1-μm TiN films both in balanced and unbalanced magnetron regimes was observed after 5 min of exposure to alkaline solution. Noticeable erosion of the surfaces took place after about 2 h of the test. That is, the corrosion resistance of samples was significantly enhanced due to protective TiN films.

3.4. Fatigue tests

Samples after fatigue tests are shown in figure 6. The sample was clamped at its thicker end, while the thinner end was subjected to the oscillating load. The load value varied in the range of 12–20 kgf/mm^2, the vibration frequency was 1680–1880 Hz. Each test procedure lasted until either the sample was fractured or after 20×10^6 cycles.

Figure 6. Samples after fatigue tests.

After the tests, the micro- and macrostructure of samples were studied in optical microscope Olympus GX51 with magnifications 100, 200, and 500. Prior to examination, the samples were grinded, mechanically polished, and etched with Keller's reagent (H_2O — 95 cm^3, HNO_3 — 2.5 cm^3, HCl — 1.5 cm^3, HF — 1 cm^3).

Samples without TiN film demonstrated the following results: one sample loaded to 20 kgf/mm^2 and two samples loaded to 16 kgf/mm^2 were broken, while all 6 samples loaded to 14 kgf/mm^2 did not fail after 20×10^6 cycles. The samples with TiN film demonstrated better performance: all 4 samples loaded to 16 kgf/mm^2 did not fail after 20×10^6 cycles, though the sample loaded to 18 kgf/mm^2 was broken.

Analyses of unbroken samples did not reveal any cracks. Broken samples had three specific regions: the region of crack initiation and stable crack growth, the region of accelerated crack growth, and the fracture area.

4. Conclusion

Magnetron-sputtered protective TiN films enhanced the surface microhardness 1.3–2 times compared to the initial value. The corrosion resistance of the surface in tests with alkali solution was significantly improved due to TiN films. The coatings demonstrated very good adhesion and showed no exfoliation after microhardness and scratch tests.

Acknowledgment

The work was supported by contracts №02.G25.31.0010 and №14.Y26.31.0008 with the Ministry of Education and Science of the Russian Federation, grant no. 16-32-00825\16 from the Russian Foundation for Basic Research and by National Research Nuclear University MEPhI in the framework of the Russian Academic Excellence Project (contract No. 02.a03.21.0005, 27.08.2013).

References

[1] Borgioli F, Galvanetto E, Galliano F P, Bacci T 2001 *Surf. Coat. Technol.* **141** 103
[2] Carter T J 2005 *Eng. Fail. Analysis* **12(2)** 237
[3] Swadźba L, Maciejny A, Formanek B, Liberski P, Podolski P, Mendala B, Gabriel H, Poznańska A 1996 *Surf. Coat. Technol.* **78** 137
[4] Bielawski M, Beres W 2007 *Wear* **262** 167
[5] Yang Q, Seo D Y, Zhao L R, Zeng X T 2004 *Surf. Coat. Technol.* **188-189** 168
[6] Creus J, Mazille H, Idrissi H 2000 *Surf. Coat. Technol.* **130** 224
[7] Liu C, Leyland A, Bi Q, Matthews A 2001 *Surf. Coat. Technol.* **141** 164

Journal of Physics: Conference Series **748** (2016) 012014 doi:10.1088/1742-6596/748/1/012014

Selection of materials for tokamak plasma facing elements based on a liquid tin capillary pore system

I E Lyublinski[1,2], A V Vertkov[1], M Yu Zharkov[1],O N Sevryukov[2], P S Dzhumaev[2],V A Shumskiy[2], A A Ivannikov[2]

[1]JSC "Red Star", Moscow, Russia
[2]National Research Nuclear University MEPhI (Moscow Engineering Physics Institute), 31 Kashirskoe highway, Moscow 115409, Russia

Corresponding author e-mail: lyublinski@yandex.ru

Abstract. Capillary-Pore Systems (CPS) filled by liquid metals are considered as an alternative solution of materials choice for plasma facing component of tokamak reactor. Tin is viewed as one of the candidates for CPS because it has lower corrosiveness than gallium and lower saturated vapour pressure compared to lithium. The corrosion resistance of Mo, Nb and W in pure liquid tin was investigated. The corrosion tests were carried out in the static isothermal conditions at a temperature up to 1050°C. As a result of the corrosion study, it was found that Mo does not corrode in liquid Sn, as opposed to Nb and is compatible with liquid tin in temperatures of up to approx. 1000°C. This allows considering Mo as an alloy base material of the in-vessel tokamak elements based on liquid tin capillary pore systems.

1. Introduction

The usage of Capillary-Pore Systems (CPS) with liquid metals is being considered as an alternative solution to the problems related to the usage of solid materials in the development of the plasma-facing elements of the Thermonuclear Neutron Source (TNS) and DEMO-like reactor [1, 2]. The main advantages of CPS as opposed to the solid materials are their resistance to property degradation and the CPS-based in-vessel component self-regeneration under conditions of normal plasma discharge, ELMs and plasma disruption. Lithium, gallium and tin are considered as the possible low melting metals for usage in tokamaks in conjunction with CPS [3–6].

The selection of liquid metal should be justified by the full analysis of the following aspects of their application: providing wettability for the realization of stability and self-regeneration of the liquid metal surface; corrosion compatibility with the CPS and tokamak materials under the conditions of normal plasma discharge, plasma instability and plasma disruption; the quantity of the liquid metal atoms entering the plasma and the effect of the atoms on the plasma parameters; the technological specifications of using these metals; technical safety; interaction of liquid metal atoms with hydrogen isotopes and the availability of the extraction technology for possible products.

2. Results

Tin is viewed as one of the candidates for CPS because it has a lower corrosiveness than gallium and a lower saturated vapor pressure compared to lithium. It was shown previously [6,7], that only refractory metals such as W, Nb, and Mo can be regarded as materials compatible with tin at high temperatures.

The research of the wettability process of W and Mo with tin has shown that Mo is effectively wetted by tin at ≥ 950°C, while W is wetted well at ≥ 1050°C (figure 1).

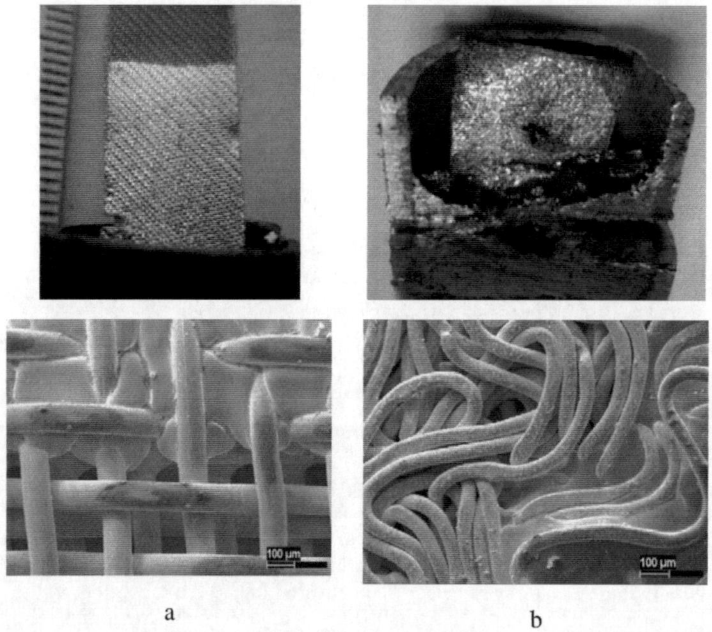

Figure 1. The wettability of metals by liquid tin in vacuum. a – Mo mesh, 950°C, 1 hour; b – W felt, 1050°C, 0.6 hours.

Corrosion of molybdenum and tungsten in liquid tin was studied by putting the sample in the Nb-2%Mo-0.1%Zr alloy capsule held at 1400°C in vacuum for 1 hour.

It is shown [8] that Mo-Sn and Nb-Sn systems can form intermetallic compounds. Mo-Sn and Nb-Sn phase diagrams (figure 2) show that at temperature of 1400°C about 3% of niobium is present as a solution in tin, whereas molybdenum is insoluble in liquid Sn even at over 2000°C.

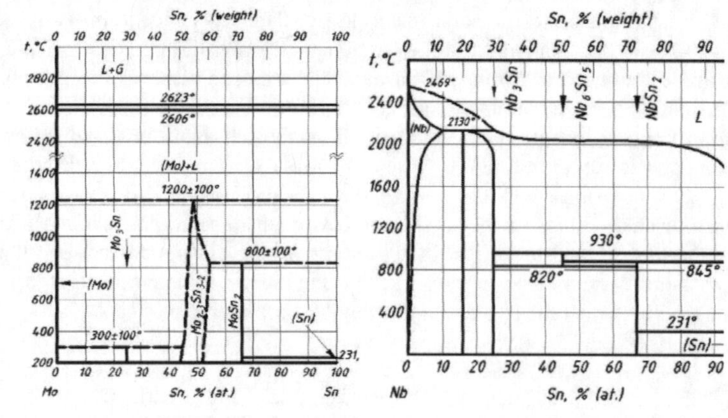

Figure 2. Mo-Sn (a), Nb-Sn (b) phase diagrams.

Journal of Physics: Conference Series **748** (2016) 012014 doi:10.1088/1742-6596/748/1/012014

Information on the interactions in the W-Sn system at temperatures under 1680°C was not found in the literature [8].

No interaction was observed on the macrophotography of the molybdenum sample after corrosiveness test with tin (figure 3). During a more thorough research changes were recorded in the near-surface layer of molybdenum sample (figure 4). The study of the distribution of elements included in the testing system (Mo, Sn, Nb) in the molybdenum near-surface layer has shown the following (see figure 5).

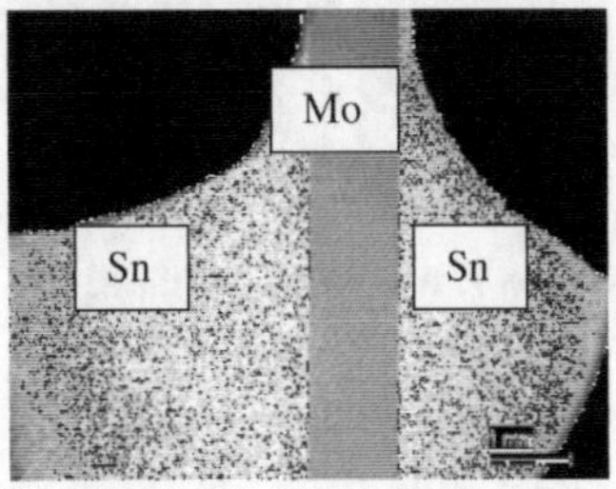

Figure 3. Macrophotography of the Mo-Sn cross-section after corrosiveness experiments.

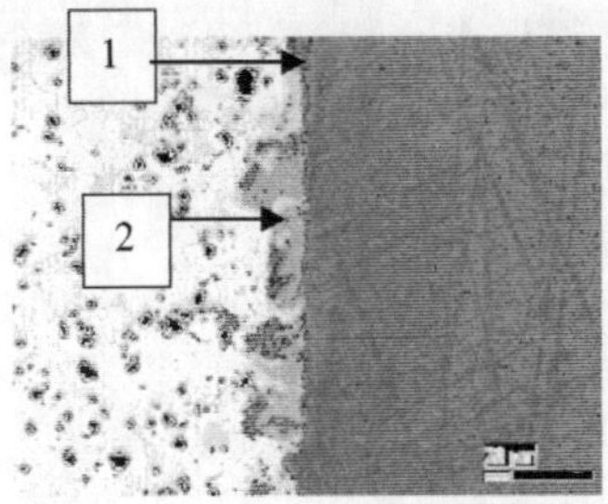

Figure 4. SEM image of the Mo-Sn interaction zone. 1 – Nb-Sn interaction zone; 2 – Nb crystals.

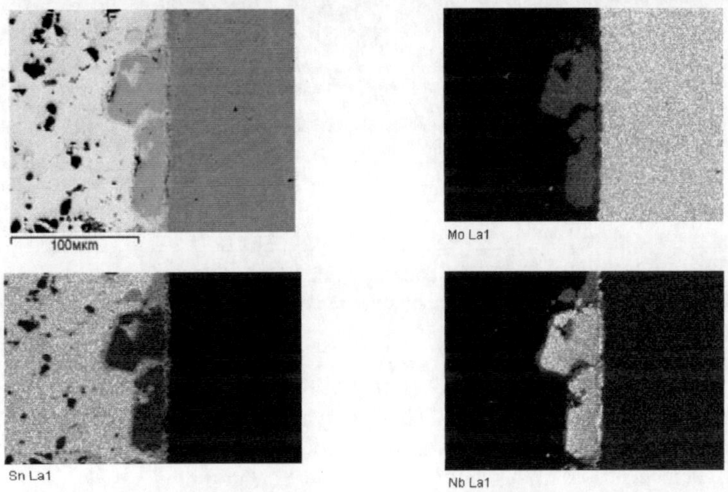

Figure 5. Microstructure and element distribution maps in the Mo-Sn interaction zone.

In the approx. 2 μm thick layer near the surface of molybdenum an enrichment of both Nb and Sn is observed. This is a product of the Nb-Sn interaction which was transferred from the surface of the Nb test capsule on the surface of the Mo sample by the process of isothermal mass transfer. Niobium crystals can be observed on the surface of molybdenum (figure 4) which precipitated from the niobium solution in tin (saturated concentration of niobium in tin at 1400° is about 3%) during the cooling.

The interaction between the niobium alloy forming a testing capsule and tin was also studied to complete the research on the material interaction.

The macrophotography of the walls of the niobium capsule in figure 6 clearly shows the interaction zone after the corrosiveness test with tin.

The detailed study of the Nb surface layer has found the existence of the interaction zone of about 20 μm thick (figure 7).

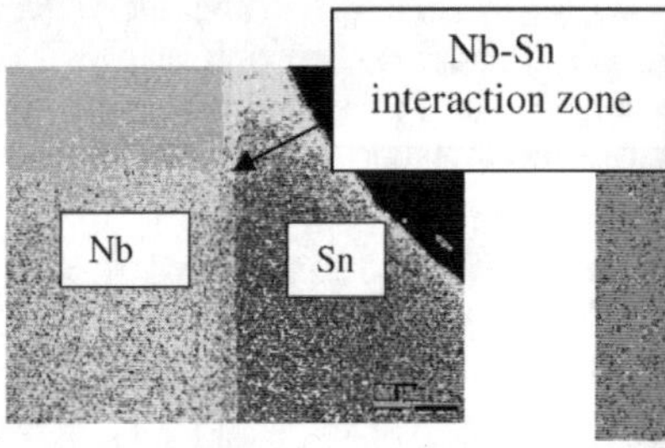

Figure 6. Macrophotography of the Nb-Sn cross-section after corrosiveness tests.

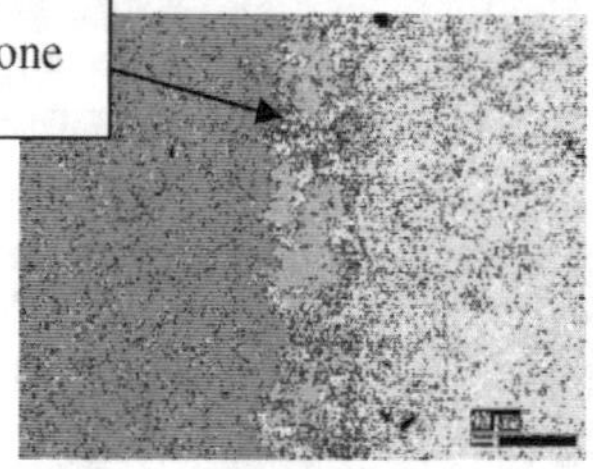

Figure 7. Nb-Sn SEM image of the Nb-Sn interaction zone.

The study of the distribution of elements included in the system undergoing tests (Mo, Nb, Sn) in the near-surface layer of the niobium capsule has shown the following (see figure 8).

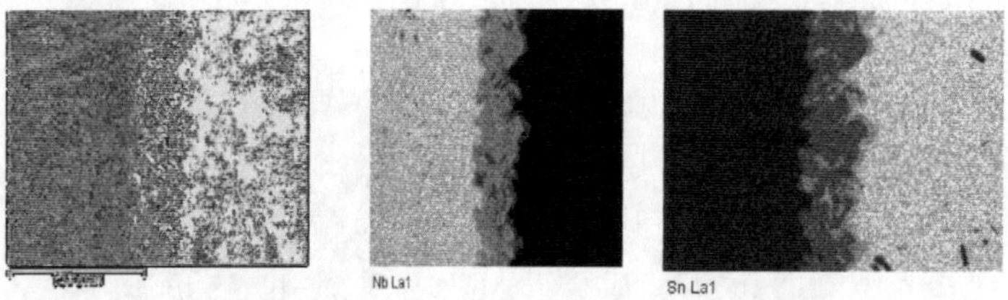

Figure 8. Microstructure and element distribution maps in the interaction zone.

The 20 μm thick interaction zone is enriched by both niobium and tin, suggesting formation of the Nb-Sn intermetallic compounds. No significant amounts of molybdenum were found in that zone.

3. Conclusion

As a result of the corrosion study, it was found that Mo does not corrode in liquid Sn, as opposed to Nb (which is furthermore excluded due to high activation), and is compatible with liquid Sn in temperatures of up to approx. 1000°C. This allows considering Mo as an alloy base material of the in-vessel tokamak elements based on liquid tin capillary pore systems.

Acknowledgments
The work was supported by Russian Science Foundation grant № 16-19-10457 and by National Research Nuclear University MEPhI in the framework of the Russian Academic Excellence Project (contract No. 02.a03.21.0005, 27.08.2013).

References

[1] Mikhaylov V N, Evtikhin V A, Lyublinski I E, Vertkov A V, Chumanov A N 1999 Lithium for Fusion Reactors and Space Nuclear Power Systems of XXI Century *(Moscow: Energoatomizdat)* (in Russian)
[2] Lyublinski I, Lazarev V, Mirnov S, Vertkov A 2015 *J. Nucl. Mater.* **463** 1156
[3] Evtikhin V A, Lyublinski I E, Vertkov A V et al. 2002 *Plasma Phys. Controlled Fusion* **44** **(6)** 955

[4] Mirnov S V, Demyanenko V N, Muravev E V 1992 *J. Nucl. Mater.* **196—198** 45
[5] Mazzitelli G, Apicella M L, Marinucci M et al. Status and Perspectives of the Liquid Material Experiments in FTU and ISTTOK *Fusion Energy* 2008 (Proc. 22th Intern. Conf. Geneva, 2008) Vienna: IAEA, 2008, CD-ROM file EX/P4-6
[6] Lyublinskiy I E, Vertkov A V, Semenov V V 2015 *PAST* **38** **(1)** 7 (in Russian)
[7] Petrova A R, Kaufman V G, Vdovina N M, Shakhnes Yu A 1969 *Metal Science and Heat Treatment* **10** 73 (in Russian)
[8] 1996-2000 *Binary Metallic System Phase Diagrams*. (Handbook in 3 vol./ Ed. acad. N P Lyakishev. Moscow: Mechanical Engineering) (in Russian)

Journal of Physics: Conference Series **748** (2016) 012015 doi:10.1088/1742-6596/748/1/012015

Analytical model of plasma-chemical etching in planar reactor

D S Veselov[1,2], A D Bakun[1], Yu A Voronov[1], V Yu Kireev[2] and O V Vasileva[1]
[1]National Research Nuclear University MEPhI (Moscow Engineering Physics Institute), 31 Kashirskoe highway, Moscow 115409, Russia
[2]National Research University of Electronic Technology, 1 Shokin Square, Zelenograd 124498, Russia

Corresponding author e-mail: dsveselov@mephi.ru

Abstract. The paper discusses an analytical model of plasma-chemical etching in planar diode-type reactor. Analytical expressions of etch rate and etch anisotropy were obtained. It is shown that etch anisotropy increases with increasing the ion current and ion energy. At the same time, etch selectivity of processed material decreases as compared with the mask. Etch rate decreases with the distance from the centre axis of the reactor. To decrease the loading effect, it is necessary to reduce the wafer temperature and pressure in the reactor, as well as increase the gas flow rate through the reactor.

1. Introduction

At the present time, technological units based on planar diode-type plasma-chemical reactors occupy the leading position in the market of plasma-chemical etching equipment. Such units are easy to operate, have low cost and can be produced both in industrial versions and in compact versions for research. However, this reactor configuration is characterized by a number of disadvantages. These include the dependence of etch parameters on the sample area (loading effect) and low etch rate compared with other systems of plasma-chemical etching [1-3]. Modeling etching in such reactors allows one to consider these disadvantages and reduce their impact on etching quality. The aim of this work is to develop an analytical model and analyze the plasma-chemical etching in the planar diode-type reactor.

2. Analytical model

This paper considers a simple analytical model of the etching process that allows us to calculate the etch rate and etch anisotropy in the planar diode-type plasma-chemical reactor. Plasma-chemical etching occurs as a result of the combined action of the following processes [4]:

- spontaneous chemical etching,
- ion-induced chemical etching,
- physical sputtering or chemically modified physical sputtering.

Typically, only one of these processes dominates the etching. The role of physical sputtering in removing material is insignificant in plasma-chemical etching. Providing high chips density on the wafer surface requires high etch anisotropy. This is only possible if ion-induced chemical reactions dominate the etching process, and also means that spontaneous chemical reactions contribute little to the etch rate. The reactor configuration used in the model is shown in figure 1.

Etch rate v and etch anisotropy A of plasma-chemical etching dominated by ion-induced reactions can be determined using the following relations [5]:

Journal of Physics: Conference Series **748** (2016) 012015 doi:10.1088/1742-6596/748/1/012015

$$v = v_{sp} + v_{in}\,, \tag{1}$$

$$A = \frac{v_{sp}+v_{in}}{v_{sp}} = 1 + \frac{v_{in}}{v_{sp}}\,, \tag{2}$$

where v_{sp} is the spontaneous etch rate and v_{in} is the ion-induced etch rate.

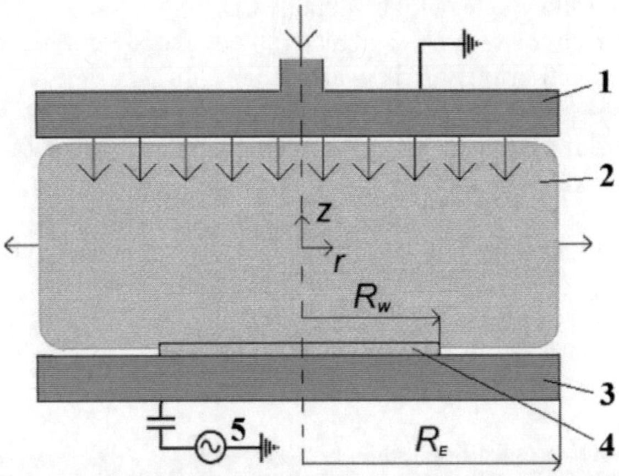

Figure 1. Planar diode-type reactor (arrows indicate the movement of operating gas): 1 - grounded counter electrode with gas diffuser; 2 - reaction volume; 3 - base electrode with radius R_E equipped with resistance matching system; 4 - wafer with radius R_W; 5 - high-frequency voltage source.

The spontaneous etch rate can be expressed as follows [6]:

$$v_{sp} = \frac{n_r \cdot Y_{sp} \cdot A_m}{\left(\frac{1}{k_{cr}}+\frac{1}{\beta}\right)\cdot N_A \cdot \rho_m} = \frac{n_r \cdot Y_{sp}}{\left(\frac{1}{k_{cr}}+\frac{1}{\beta}\right)\cdot \rho_a}\,, \tag{3}$$

where n_r is the concentration of chemically active particles in the reaction volume, k_{cr} is the etching reaction rate constant, β is the mass-transfer coefficient of chemically active particles equal to the ratio of the particles diffusion coefficient D_r and the thickness δ of the boundary layer near the surface of processed material in which the concentration of chemically active particles decreases from its reaction volume value to the concentration n_{rs} on the surface of processed material, Y_{sp} is the coefficient of spontaneous etching of processed material that equals the number of chemically active particles required to remove one atom of material, A_m is the atomic mass of processed material, N_A is the Avogadro constant, ρ_m is the mass density of processed material and ρ_a is the atomic density of processed material, which can be expressed as:

$$\rho_a = \rho_m \cdot \frac{N_A}{A_m}\,. \tag{4}$$

The character of spontaneous etching is determined by the ratio between k_{cr} and β. If $\beta \gg k_{cr}$, the spontaneous etch rate of the processed material is the same as the chemical reaction rate on its surface. In this case, the concentration of chemically active particles on the surface of the material is the same as the concentration of chemically active particles in the reaction volume. Such processes are limited by the kinetics of the chemical reaction, i.e., occur in the kinetic region. Their etch rate v_{spk} is determined by the activation energy of the chemical reaction, i.e., by the etchant composition, and can be expressed as:

$$v_{spk} = n_r(r) \cdot \frac{Y_{sp} \cdot k_{cr}}{\rho_a}\,. \tag{5}$$

If, on the contrary, $k_{cr} \gg \beta$, the spontaneous etch rate of the processed material is determined by the diffusion rate of chemically active particles moving toward its surface. The concentration of

Journal of Physics: Conference Series **748** (2016) 012015 doi:10.1088/1742-6596/748/1/012015

chemically active particles near the surface of the material is much less than the concentration of chemically active particles in the reaction volume. Such processes are limited by diffusion of chemically active particles, i.e., occur in the diffusion region. In the diffusion region, all reactions are not dependent on the specific features of the chemical mechanism of material etching. In this region, the etch rates are only determined by the diffusion coefficients of the used chemically active particles. In the diffusion region, the following cannot be obtained:

- high values of etch selectivity of processed material compared with masking layer;
- high values of etch rate uniformity on the wafer area.

For this reason, plasma-chemical etching in planar reactors should not be performed in diffusion region. This region is characterized by the high pressure of the gas mixture, high temperature of processed wafers and low gas flow rate through the reactor since the boundary layer thickness δ is inversely related to the square root of gas flow rate [6]. The practical demonstration of the disadvantages of plasma-chemical etching in diffusion region is the loading effect.

The etch rate due to ion-induced reactions is directly proportional to the density of the ion current $j_i(r)$. For singly charged ions, it can be expressed as follows:

$$v_{in} = \frac{k_{in}(E_i)\cdot j_i(r)\cdot Y_{in}}{\rho_a\cdot e}, \tag{6}$$

where $k_{in}(E_i)$ is the constant of ion-induced chemical reaction of material etching that depends on ion energy E_i, Y_{in} is the coefficient of ion-induced etching equal to the number of ions required to remove one atom of the material and e is the ion charge [7].

Substituting expressions (5) and (6) into equations (1) and (2), we obtain the following expressions for the rate and anisotropy of plasma-chemical etching corresponding to spontaneous etching in the kinetic region:

$$v = n_r(r)\cdot\frac{Y_{sp}\cdot k_{cr}}{\rho_a} + \frac{k_{in}(E_i)\cdot j_i(r)\cdot Y_{in}}{\rho_a\cdot e}, \tag{7}$$

$$A = 1 + \frac{k_{in}(E_i)\cdot j_i(r)\cdot Y_{in}}{e\cdot n_r(r)\cdot Y_{sp}\cdot k_{cr}(T_m)}. \tag{8}$$

Expression (7) shows that the etch rate and uniformity are determined by the level and distribution of the processed material temperature T_m. The etching process is also influenced by the concentration of chemically active particles and by the density of ion current. The concentration of chemically active particles in the reaction volume depends on the following parameters of the reactor [8]:

- the ratio between gas flow rate and diffusion rate of chemically active particles;
- the ratio between volumetric generation rate of chemically active particles and their diffusion rate;
- the ratio between conversion time of operating gas molecules into chemically active particles and the lifetime of the molecules in the reaction volume.

To provide a constant concentration of chemically active particles in the reaction volume, it is necessary to gradually reduce the flow rate of the operating gas Q_g during plasma-chemical etching. However, the flow rate of etch products Q_{pr} imposes a limitation on the operating gas flow rate reduction. At high rate of etch products flow, material etching with low uniformity occurs on the wafer surface. Deposition of etch products on the wafer surface and reactor walls is also possible. Therefore, reducing the flow rate of operating gas should comply with the following condition [4]:

$$Q_{pr} \le 0{,}1\cdot Q_g, \tag{9}$$

in which the flow rate of etch products is calculated in accordance with the formula:

$$Q_{pr} = 2.23\cdot 10^{-25}\rho_a\cdot S_m\cdot v, \tag{10}$$

where S_m is the etch area on the surface of the processed wafer.

Thus, when condition (9) is fulfilled in the kinetic region, the concentration of chemically active particles in the reaction volume is independent of the radial coordinate r, meaning that in equations (7) and (8) $n_r(r) = n_r = const$. This conclusion is correct if we neglect recombination of chemically active particles and at their low mean free path in comparison with the distance between the electrodes which is true at the operating gas pressure in the reactor less than 100 Pa. If the thermal contact between the wafer and the electrode is good, the wafer temperature is independent of its radius, i.e., $T_m(r) = T_m = const$. In this case, the spontaneous etch rate is constant on the wafer area.

To determine the ion-induced etch rate, it is important to know the ion current. In case of ambipolar diffusion, the dependence of the ion current density on the planar reactor radius can be expressed as follows [4]:

$$j_i(r) = j_i(0) \cdot J_0\left(2.4 \cdot \frac{r}{R_E}\right), \tag{11}$$

where J_0 is the Bessel function of zero order.

Consequently, the ion-induced etch rate in the centre of the electrode is twice higher than at three-quarters of its radius. On the edge of the electrode, ion-induced etch rate is zero. To reduce the influence of the ion-induced etch rate, electrodes with a radius that is significantly bigger than that of the processed wafer are usually chosen. At half radius, the ion-induced etch rate is about 70% of its maximum value. At this point, the nonuniformity of resultant etch rate decreases due to spontaneous etching. To improve the uniformity of the etch rate, planar reactors with reduced distance between electrodes can be used. In addition, the peripheral region's resistance to high-frequency field can be reduced using dielectric overlays.

Equation (8) shows that to obtain simultaneous uniformity of etch rate and etch anisotropy of plasma-chemical etching on the wafer area (on the electrode area of the planar reactor) it is necessary to provide uniformity of both spontaneous and ion-induced etching. Expressions presented in the paper are valid for etching of any material. The selectivity of plasma-chemical etching of various materials is determined by the difference of activation energies of their respective chemical reactions. The higher the activation energy of chemical reaction, the lower the etch rate. The activation energy of chemical reaction of the mask material needs to be much higher than that of the processed material. However, it should be taken into account that when increasing the ion current density and ion energy, the etch rate of various materials changes differently. The etch rate of materials with a lower value of activation energy of chemical reaction increases slower than that of materials with a higher value of activation energy. Consequently, with increasing the ion current and ion energy, the difference of etch rate between various materials (etch selectivity) decreases.

3. Conclusions

To provide anisotropic plasma-chemical etching of micro- and nanostructural elements in the planar diode-type reactor, simultaneous execution of a large number of sometimes contradictory conditions is required. High anisotropy occurs when the ion-induced etch rate is significantly higher than that of the spontaneous etch rate. This requires high ion energy and high values of ion current that also lead to the increase in wafer temperature, which is undesirable. Moreover, this reduces etch selectivity of various materials (e.g., mask and processed material). Spontaneous etching should take place in the kinetic region. For this purpose, it is necessary to gradually reduce the flow rate of operating gas and the pressure of gas mixture in the reactor in the process of plasma-chemical etching. To provide the uniformity of the etch rate on the wafer area, the wafer diameter should be no more than three-quarters that of the electrode diameter. These are general recommendations. For each specific objective, it is necessary to take into account the specified etching parameters when constructing a reactor.

Acknowledgements

This work was supported by National Research Nuclear University MEPhI in the framework of the Russian Academic Excellence Project (contract No. 02.a03.21.0005, 27.08.2013).

References

[1] Raton B 2008 *Handbook of Semiconductor Manufacturing Technology* (Second Edition)

Journal of Physics: Conference Series **748** (2016) 012015 doi:10.1088/1742-6596/748/1/012015

Nishi Y and Doering R (New York: Taylor & Francis Group)
[2] Hopfe V and Sheel D W 2007 *Plasma Proc. Polym.* 4 **3** 253–265
[3] Otto T, Wolf H, Streiter R, Dehoff A, Wandel K and Gessner T 1999 *Microelectron. Eng.* 45 **4** 377–391
[4] Kireev V Yu 1999 *Engineering J.* **2(23)** 58 (in Russian)
[5] Danilin B S and Kireev V Yu 1987 *Using low-temperature plasma for etching and cleaning materials* (Moscow: Energoatomisdat) (in Russian)
[6] Kireev V Yu, Danilin B S and Kuznetsov V I 1983 *Plasma-chemical and ion-chemical etching of microstructures* (Moscow: Radio i sviaz) (in Russian)
[7] Jansen H Boer M Legtenberg R and Elwenspoek M 1995 *J. Micromech. Microeng.* **5(2)** 115
[8] Levenspiel O 1972 *Chemical reaction engineering* (New York: John Wiley & Sons)

Facility and the method for MEIS analysis of layers redeposited in plasma devices

D Bulgadaryan, D Kolodko, V Kurnaev and D Sinelnikov

National Research Nuclear University MEPhI (Moscow Engineering Physics Institute), 31 Kashirskoe highway, Moscow 115409, Russia

E-mail: DGBulgadaryan@mephi.ru

Abstract. The experimental facility for surface analysis by means of middle energy ion scattering (MEIS) is described. Experimental results are compared with SCATTER computer code simulation. The simulation shows the possibility to determine the thickness of very thin redeposited layers of heavy elements by keV energy hydrogen ions scattering spectroscopy.

1. Introduction

In fusion devices, an interaction between plasma and the first wall leads to erosion of plasma-facing materials that can result in deposition of thin layers of sputtered materials. Therefore, locating redeposited layers and determining their thickness are two acute problems. Various means of surface analysis are used for this purpose. However, methods using secondary emission spectroscopy, such as SIMS, do not have sufficient depth resolution and, as a result, they destroy a surface during the analysis. Rutherford backscattering spectroscopy (RBS) [1] or spectroscopy of reflected electrons (SRE) [2] can be also used as non-destructive analytical methods, but their depth resolution is not good enough for very thin layers. Besides, RBS requires the use of big/expensive accelerators, while the energy analyzer with very high absolute resolution is required for SRE. At the same time, it was shown earlier [3] that middle energy ion scattering (MEIS) technique can be applied as a non-destructive method of thin layers analysis with high depth resolution (~ 0.3nm). In this work we describe the experimental setup for application of MEIS for analysis of plasma-facing materials.

2. Experimental setup

Experimental scheme is shown in figure 1 (a). A duoplasmatron is used as an ion source with the automated power supply and gas flow control. The ion source hot cathode current and discharge parameters are controlled by the power supplies with analog control and Spellman UM40P30 power supply with the output voltage $U = 1\div 40$ kV is used for ion beam acceleration.

The separating electromagnet with main radius for an ion trajectory, $R = 50$ cm, at maximum magnetic field of ~ 0.4 T, is capable to separate the ions with the mass to charge ratio (M/Z) at the accelerating voltage, U, satisfying the relation (M(a.m.u.)$/Z$)$\cdot U$(eV) $\leq 2\cdot 10^6$. Therefore, single charged Ar^+ ions at maximum energy of 40 keV can be separated. The electromagnet is powered by TDK-Lambda Genesys 30-50 supply unit with the computer control via RS-232 protocol. The monochromaticity of an ion beam at 5 keV is ~ $5\cdot 10^{-3}$, and its angular divergence is ~0.01 rad [4]. The beam density can be controlled with 3 electrode einzel lens of the ion-optical system.

The interaction chamber is pumped by the turbomolecular pump with an output of 800 l/s that allows obtaining the pressure of ~10^{-6} Torr during the target heating and when the ion beam is on.

Residual gas pressure after a long term pumping and outgassing is ~$2{\cdot}10^{-8}$ Torr. An additional quadrupole mass spectrometer and an ion source for target surface modification can be installed on the chamber. Targets can be up to 20 mm in diameter and are placed on the rotating goniometer that allows irradiation at different angles of beam incidence. The target is heated by the halogen lamp to the temperatures of about 400 °C; the temperature is measured with a chromium-aluminum thermocouple connected to the Aktakom AM-1152 multimeter. Deflecting plates can be used for the additional focusing of the beam onto the target. The ion current to the target is controlled by the Keithley 6485 picoammeter and recorded during the experiment. The ion current is used for determination of the ion fluence on the target.

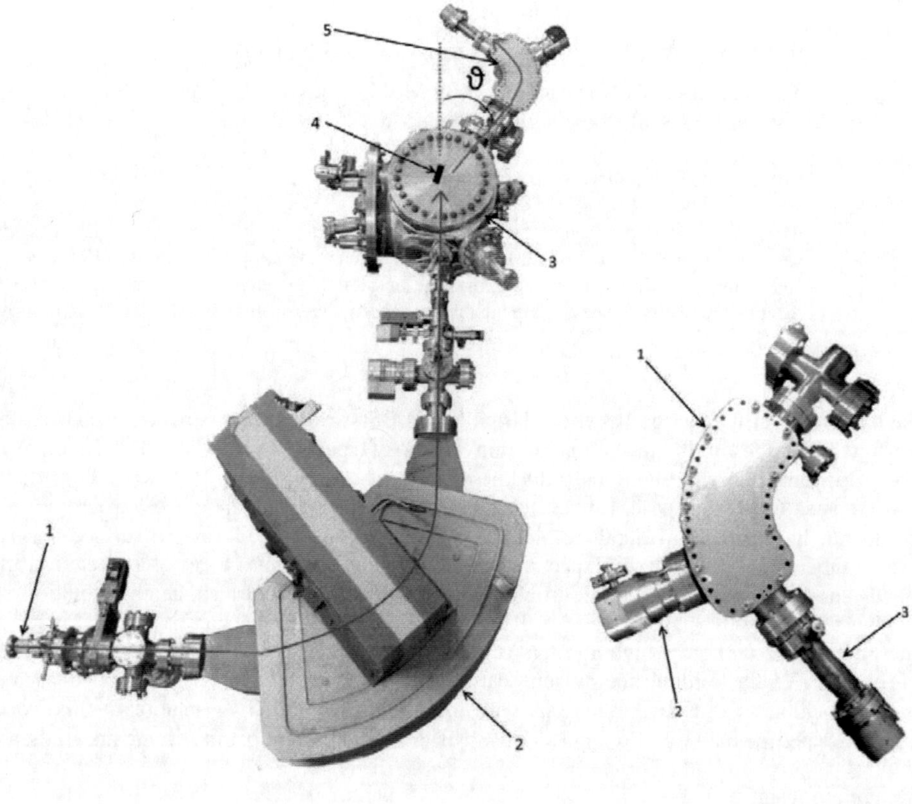

a) b)

Figure 1. a) experimental setup: 1 – ion source, 2 – electromagnet, 3 – interaction chamber, 4 – target, 5 – energy analyzer, red line – primary beam direction, green line – scattered beam direction, θ – detection angle; b) energy analyzing system: 1 – quarter-spherical deflector, 2 – Pfeiffer TMU 071 turbomolecular pump, 3 – VEU-1A secondary electron multiplier.

A quarter-spherical deflector with 21 cm radius for the central trajectory (figure 1 (b)) is assembled and tested as an energy analyzer to implement the technique of ion scattering and recoil ion spectroscopy. Separate pumping system of the analyzer allows measurements of scattered ions even at high gas pressure at interaction chamber during the target treatment with gas, plasma and ions. At a solid angle of scattered particle detecting of $2.1{\cdot}10^{-4}$sr, with the half-width of the scattering angle $\Delta\theta = \pm 0.5°$, its energy resolution is not larger than 1%. Power supply system of the analyzer consists of the Keithley 6517B electrometer used as a precise voltage source to control Advanced HiVolt PSM10 units that are connected to the plates of the analyzer through high voltage dividers. The system allows separating charged particles in steps of 12 eV. The analyzer records scattered and recoil ions at angle θ = 37° relative to the primary ion beam direction. Ion current is recorded with VEU-1A secondary electron multiplier and the Keithley 6517B electrometer connected to PC.

The experimental setup is computer-controlled. The control is carried out using LabVIEW software, RS-232 protocol and ADC-DAC ICP-CON modules for connection with analog-controlled supply units.

3. Results and discussion

The results of the surface analysis of W target with the adsorbed water layers by using 17300 eV Ar^+ beam are shown in figure 2 (a). Scattering angle is θ = 37°, the graphs are normalized to the initial beam energy and signal intensity. Small peaks on the main graph correspond to recoil H^+ an O^+ ions. Spectroscopy of negative ions knocked out from a target by primary heavy ion beam can be used for elemental analysis of adsorbed surface layers (O^- - additional graph). Figure 2 (b) shows the experimental and simulated spectra of H^+ with the initial energy of 8600 eV scattered on carbon and tungsten surfaces at angle θ = 37° relative to the initial direction of beam incidence, normalized to the initial beam energy. SCATTER pair collision computer code [5] was used to simulate the ion scattering processes with the same parameters. The comparison between experimental and simulation results shows a good agreement. Determination of the charged fraction was conducted using the semi-empirical formula [6] for hydrogen scattering:

$$\eta_0^+ = kE^{\mu} \tag{1}$$

where $k \approx 0.08$, $\mu \approx 0.5$, E is expressed in keV.

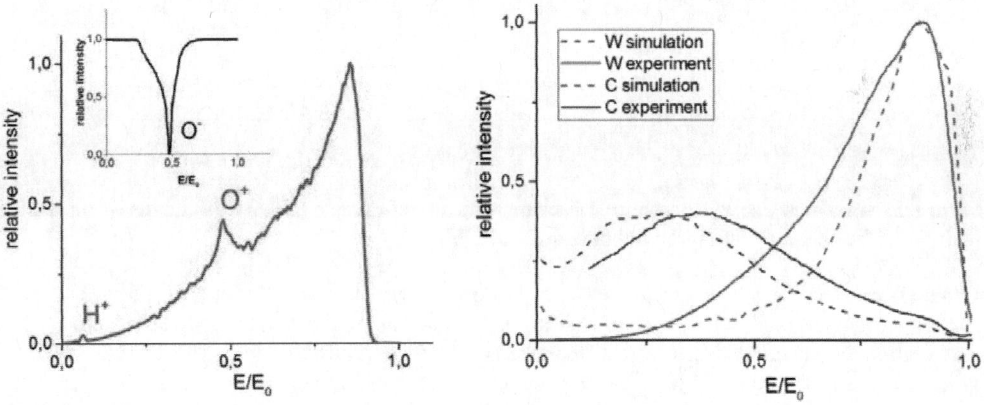

a) b)

Figure 2. a) energy spectra of Ar^+ with initial energy E_0 = 17300 eV scattered on W target with adsorbed water surface layers with peaks of recoil ions; b) energy spectra of H^+ with initial energy E_0 = 8600 eV scattered on bulk graphite and tungsten targets.

Computer modeling using SCATTER code was performed to simulate possible evolution of energy spectra during deposition of heavy element or erosion of marker target under plasma impact. Boron carbide (B_4C) is proposed to be used as one of ITER materials [7], so the scattering of hydrogen keV energy ions on boron target (since the average molecular mass of B_4C is almost the same as that of B) with thin surface layer of tungsten for different W thickness is considered as example. Detection angle of scattered particles is $\theta = 37^{\circ}$ relative to the primary ion beam direction and corresponds to the experimental setup. The number of particles used in each calculation is 10^7.

The spectra of scattered H^+ beam with energy of 10 keV depending on the surface W layer thickness are shown in figure 3. The increase of W layer thickness results in broadening of W peak and slight increment of its amplitude. The FWHM grows rapidly from 0.03 for 0.5 nm thick W layer to 0.14 for 2 nm thick W layer. For 4 and 5 nm thick W layers the FWHM is 0.23 and 0.25, respectively, which is relatively close to the bulk tungsten (FWHM = 0.29). Therefore, the width and amplitude of high-energy peaks obtained with keV hydrogen MEIS can be used to determine the thickness of very thin layers of heavy element (like W) - from a few angstroms up to ~3 nm, deposited on light substrate in plasma devices. Growth of the primary ion energy and the scattering angle increases accuracy for determination of thickness of redeposited tungsten layers.

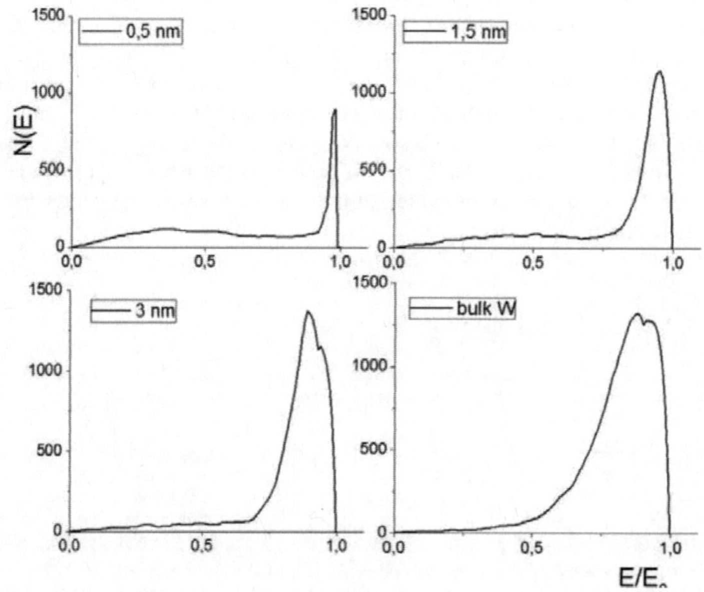

Figure 3. Simulated energy spectra of H^+ beam with initial energy E_0=10 keV scattered on B target with W surface layer of different thickness and bulk W.

4. Conclusions

The experimental facility for use of MEIS technique for analysis and control of erosion and redeposition of materials with very different atomic number is described. The energy spectra of hydrogen ions scattered on light and heavy targets are in good agreement with computer simulation results. The simulation also shows a possibility to determine the presence and thickness of very thin layers of heavy elements, like W, redeposited on light substrate by applying scattering of hydrogen ions of keV energies.

Acknowledgements

This work is partly supported by contract H.4x.44.9Б.16.1018 of SC Rosatom and by National Research Nuclear University MEPhI in the framework of the Russian Academic Excellence Project (contract No. 02.a03.21.0005, 27.08.2013).

References

[1] Krat S.et al. 2015 *J. Nucl. Mat.* **456** 106

[2] Afanas'ev V P et al. 2013 *J. Surface Investigation. X-Ray, Synchrotron and Neutron Techniques,***7(1)** 123
[3] Kurnaev V A et al. 1999 *Tech. Phys. Lett.* **25**(11) 442
[4] Mamedov N V et al. 2012 *Bull. Rus. Acad. Sci.: Physics* **76** (6) 683
[5] Trifonov N N et al. 2008 *J. Surface Investigation. X-Ray, Synchrotron and Neutron Techniques* **2** (2) 175
[6] Kurnaev V, Protasov Yu, Tsvetkov I. 2008 *Introduction to beam electronics* (Moscow: MEPhI) (in Russian)
[7] Begrambekov L et al. 2015 *J. Nucl. Mat.* **463** 792

Reactive ion etching of silicon using low-power plasma etcher

D S Veselov, A D Bakun and Yu A Voronov
National Research Nuclear University MEPhI (Moscow Engineering Physics Institute), 31 Kashirskoe highway, Moscow 115409, Russia

Corresponding author e-mail: dsveselov@mephi.ru

Abstract. The paper is devoted to the study of deep reactive ion etching of silicon using diode plasma etcher system with a low-power source. Silicon wafers were etched in a sulfur hexafluoride plasma and sulfur hexafluoride/oxygen plasma. The maximum achieved silicon etch rate was about 2 μm/min. The expediency of using dry reactive ion etching in combination with wet anisotropic etching of silicon for manufacturing of microelectromechanical systems (MEMS) was demonstrated.

1. Introduction

Vertical etching of silicon is widely used for forming relief structures in MEMS technologies. Traditionally, wet anisotropic etching of silicon is used for this purpose [1]. Its advantages are the ease of operation and low cost. Wet anisotropic etching of silicon also provides high reproducibility of processing modes and the best uniformity of etched surface [2]. Unfortunately, the etching ability of wet etchants decreases during the etching process. This leads to reduction of etch rate and uniformity of the etched surface. Therefore, forming deep cavities in silicon wafers using wet anisotropic etching often requires replacement of etchant. This complicates the production process and reduces the percentage yield.

Further development of MEMS technologies and associated technological equipment leads to increasingly frequent application of dry etching methods. Dry etching of silicon is characterized by some advantages over wet etching. First, dry etching of silicon provides near-vertical sidewalls of the etched cavities [3]. This allows one to increase the quantity of chips on the substrate, thereby reducing their production cost. Dry etching also provides higher etch rate of silicon in comparison with wet etching. In addition, the dry etching rate is less dependent on the silicon crystallographic orientation [4]. However, the application of dry etching requires expensive equipment.

MEMS technologies often require deep silicon etching with high uniformity of the etched surface. In this case, it is convenient to use dry reactive ion etching in combination with wet etching. One of the MEMS technologies is manufacturing dielectric membrane structures for sensitive elements of semiconductor gas sensors. Such structures are fabricated by dielectric film deposition on the surface of silicon substrate and subsequent substrate etching from the reverse side up to form reach-through cavities under the membranes. Initially, deep wet anisotropic etching of silicon is conducted. Then, dielectric film and sensitive layers are formed on the etched film. Finally, deep dry vertical etching of the residual silicon layer is performed to form reach-through cavities under the deposited membranes [5–7]. This allows one to exclude the influence of etchant solution on the formed sensitive layers, increasing the percentage yield. We propose the use of reactive ion etching of silicon as the dry vertical etching method in the described technology. The paper presents study results of deep reactive ion etching of silicon and expediency assessment of its use in combination with wet anisotropic etching of silicon for manufacture of MEMS based on dielectric membrane structures.

2. Study of deep reactive ion etching of silicon

Phosphorus-doped single crystal silicon wafers of orientation (100) with the resistivity of 4.5 Ωcm were used to study deep reactive ion etching of silicon. 0.5-μm-thick silicon dioxide and 0.3-μm-thick nickel films were used as etching masks. Silicon dioxide films were formed by thermal oxidation of silicon in the dry-wet-dry mode at the temperature of 1150 °C. Initially, oxidation of silicon wafers was conducted in the ambient of dry oxygen, then in the ambient of water vapor. The final stage of silicon oxidation was performed in the ambient of dry oxygen. Nickel films were formed using magnetron sputtering. The masking pattern in the form of square holes with the size of 1×1 mm^2 was obtained using photolithography and subsequent wet etching. After this, silicon dioxide films were etched in buffered oxide etchant at the temperature of 20 °C. Nickel films were etched in 25% aqueous hydrogen nitrate solution of at the temperature of 70 °C.

The study was performed using the reactive ion etcher with planar diode-type reactor. The unit configuration is schematically shown in figure 1. Silicon wafers were placed without fixing on the stationary base electrode equipped with a flowing water cooling system. The ion etcher was equipped with a single stage vacuum system. Vacuum was provided by the Agilent Varian SD-700 dual stage rotary vane vacuum pump. The base electrode was powered by the high-frequency Dressler Cesar 136 generator. The study was carried out at plasma discharge power of 50–140 W. The reaction chamber was equipped with two-channel gas injection system. The working gas was supplied into the reaction chamber through the counter electrode equipped with the gas diffuser. Gas flows into the chamber were controlled by electronic regulators of gas rate. Only high-purity gases were used in this study.

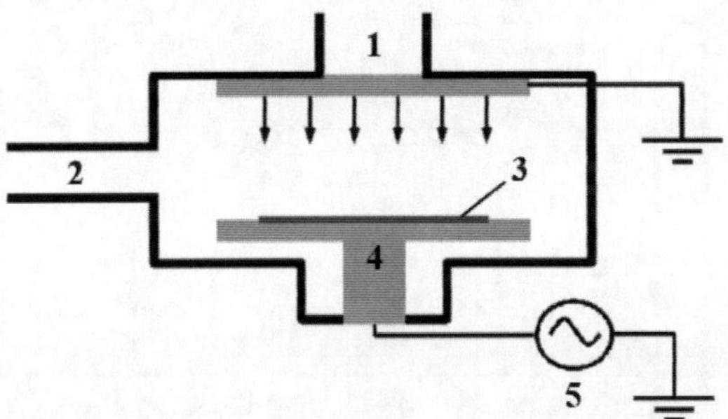

Figure 1. Reactive ion etcher: 1 - grounded counter electrode with gas diffuser; 2 - vacuum pump tract; 3 - silicon wafer; 4 - base electrode with resistance matching system; 5 - high-frequency voltage source.

Deep reactive ion etching of silicon was studied in a sulfur hexafluoride plasma and sulfur hexafluoride/oxygen plasma. The etching time of all samples in all processing modes was under 10 minutes. The aim of the study was to select the optimal silicon etching modes. Silicon etch rate was calculated from the depth of the etched cavities measured using Veeco Dektak 150 profilometer. The influence of the high-frequency power supply power on silicon etch rate was studied. The influence of sulfur hexafluoride flow rate into the reaction chamber on silicon etch rate was also studied. Dependences of the silicon etch rate on the power of the high-frequency power supply at various flow rates of sulfur hexafluoride were obtained (see figure 2). The influence of oxygen content in the mixture with sulfur hexafluoride on silicon etch rate was also studied. The flow of sulfur hexafluoride into the reaction chamber was assumed constant and equal to 18 cm^3/min. Oxygen flow was varied during the study and dependencies of the silicon etch rate on the oxygen content in the mixture with sulfur hexafluoride at various power values of the high-frequency power supply were obtained (see figure 3). The smooth curves in figures 2 and 3 passing through experimental points were obtained with interpolation by the method of cubic spline using Origin Pro 9.1 software (Origin Lab corp.).

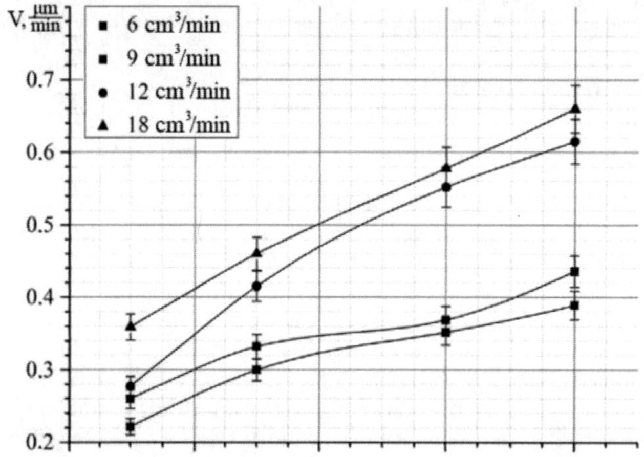

Figure 2. Silicon etch rate versus RF generator power at various sulfur hexafluoride flow rates.

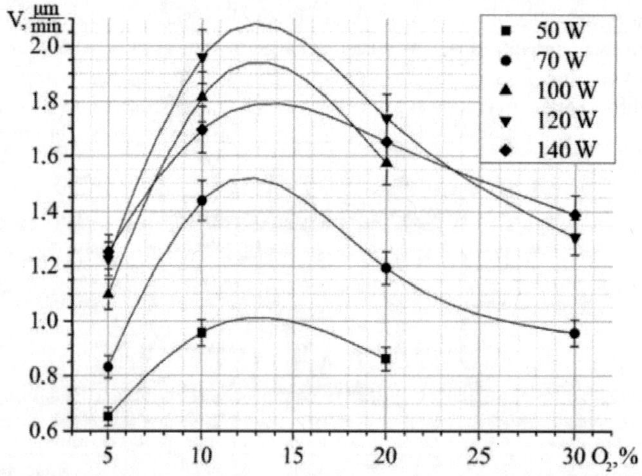

Figure 3. Silicon etch rate versus oxygen content in the gas mixture at various RF generator power values.

3. Results and discussion

As can be seen from figure 2, the silicon etch rate gradually increases with the increase in the power of the high-frequency power supply and sulfur hexafluoride flow rate into the reaction chamber. The maximum achieved silicon etch rate in the sulfur hexafluoride plasma was about 0.6 µm/min. This etch rate was obtained when the sulfur hexafluoride flow rate into the reaction chamber was of 18 cm^3/min and the power of the high-frequency power supply was of 120 W. Adding oxygen into the sulfur hexafluoride flow was seen to increase the silicon etch rate. The maximum achieved silicon etch rate in the sulfur hexafluoride/oxygen plasma was about 2 µm/min. This etch rate was obtained when the oxygen content in the sulfur hexafluoride flow was about 10%. Taking into account the mathematical approximation, the maximum silicon etch rate is achieved when the oxygen content in the sulfur hexafluoride flow is about 14%. Further increase in the oxygen content in the mixture leads to the reduction of silicon etching rate, as shown in figure 3.

In order to explain this, it is necessary to consider the processes occurring in the reaction chamber during the reactive ion etching of silicon. Plasma discharge activates successive dissociation of sulfur hexafluoride resulting in the formation of fluorine atoms that have one unpaired electron. These atoms are adsorbed by the wafer surface and react with the silicon. The product of this chemical reaction is silicon tetrafluoride which is then desorbed from the wafer surface. When feeding the mixture of sulfur hexafluoride and oxygen into reaction chamber, the following processes also occur. The plasma discharge activates oxygen dissociation in addition to the sulfur hexafluoride dissociation which leads to formation of oxygen atoms that have one unpaired electron. These oxygen atoms clean the wafer surface of the deposited sulfur, forming gaseous sulfur oxide. The oxygen atoms are also adsorbed by the wafer surface and react with silicon. As a result, a thin film of silicon oxide forms on the wafer surface.

The increase in silicon etch rate at low oxygen content in the mixture with sulfur hexafluoride is caused by the cleaning of the wafer surface through sulfur oxidation that is transferred into the gaseous state. Further increase in the oxygen content in the mixture leads to increase in silicon oxidation rate. If silicon oxide film forms on the wafer surface, fluorine ions are expended on its removal. Therefore, the resultant etch rate of silicon is reduced.

The maximum rate of reactive ion etching of silicon was obtained when the power of the high-frequency power source was of 120 W. Further increase in plasma discharge power significantly reduces the silicon etch rate. This is explained by the sputtering of the base electrode and chamber walls taking place at higher discharge power. The sputtered materials are partially deposited on the wafer surface and result in masking the silicon etching surface.

The studied silicon etching modes revealed that the 0.5-µm-thick silicon dioxide mask is mostly etched off the wafer surface. Therefore, such a mask can only be used when silicon etching is performed on the depth of several tens of micrometers. In this study, the nickel mask was characterized by high etch resistance both in the sulfur hexafluoride plasma and in the sulfur hexafluoride/oxygen plasma. This means that for deep reactive ion etching of silicon it is necessary to use metal masking layers, in particular, nickel masks.

4. Conclusions

A mode of deep reactive ion etching of silicon was found which provides sufficiently high silicon etch rate and acceptable uniformity of the etched surface. Feeding sulfur hexafluoride flow with oxygen content of 10% into the reaction chamber resulted in the silicon etch rate of about 2 µm/min. This mode allows etching 100-µm-thick silicon layers using a 0.3-µm-thick nickel mask. It was also shown that it is expedient to use dry reactive ion etching in combination with wet etching to form deep relief in silicon wafers for manufacture of MEMS. The use of deep reactive ion etching of silicon at the final stage of dielectric membrane structures formation for sensitive elements of semiconductor gas sensors improves the percentage yield. This is achieved through the absence of etchant contact with sensitive layers, and also through the absence of rapid changes in temperature and pressure typical for wet silicon etching at the final stage of the membranes formation.

Acknowledgements

This work was supported by National Research Nuclear University MEPhI in the framework of the Russian Academic Excellence Project (contract No. 02.a03.21.0005, 27.08.2013).

References

[1] Rouhi J, Mahmud S, Hutagalung S D, Naderi N, Kakooei S and Abdullah M J 2012 *Semiconductor Sci. Tech.* **6** **27** 065001
[2] Shikida M, Inagaki N, Sasaki H, Amakawa H, Fukuzawa K and Sato K 2010 *J. Micromech. Microeng.* **1** **20** 015038
[3] Azimi S, Sandoughsaz A, Amirsolaimani B, Naghsh-Nilchi J and Mohajerzadeh S 2011 *J. Micromech. Microeng.* **7** **21** 074005
[4] NozawaY 2010 *J. Vac. Soc. Japan* **7** **53** 446
[5] Veselov D S 2015 *Phys. Procedia* **72** 500
[6] Veselov D S and Voronov Yu A 2015 *Phys. Procedia* **71** 423
[7] Veselov D S, Voronov Yu A and Vanyukhin K D 2015 *Phys. Procedia* **72** 495

Journal of Physics: Conference Series **748** (2016) 012018 doi:10.1088/1742-6596/748/1/012018

Study of electrochemical properties of thin film materials obtained using plasma technologies for production of electrodes for pacemakers

O I Obrezkov[1], V P Vinogradov[1], V I Krauz[1], D V Mozgrin[2], I A Guseva[3], E S Andreev[4], A A Zverev[4], A L Starostin[5]

[1]NRC «Kurchatov Institute», 1 Kurchatov sq., Moscow 123098, Russia
[2]National Research Nuclear University MEPhI (Moscow Engineering Physics Institute), 31 Kashirskoe highway, Moscow 115409, Russia
[3]INNOVATECHPROM Ltd, 10 Ryazanskiy av., Moscow 109428, Russia
[4]NPP Bulat Ltd., 18 Komitetsky les st., Korolev 141067, Moscow region, Russia
[5]Cardioelectronics Ltd, 2 Zavodskaya st., Klimovsk 142181, Moscow region, Russia

Corresponding author e-mail: olio25845@mail.ru

Abstract. Studies of thin film materials (TFM) as coatings of tips of pacemaker electrodes implanted into the human heart have been performed. TFM coatings were deposited in vacuum by arc magnetron discharge plasma, by pulsed discharge of "Plasma Focus", and by electron beam evaporation. Simulation of electric charge transfer to the heart in physiological blood-imitator solution and determination of electrochemical properties of the coatings were carried out. TFM of highly developed surface of contact with tissue was produced by argon plasma spraying of titanium powder with subsequent coating by titanium nitride in vacuum arc assisted by Ti ion implantation. The TFM coatings of pacemaker electrode have passed necessary clinical tests and were used in medical practice. They provide low voltage myocardium stimulation thresholds within the required operating time.

Electric cardiac pacemakers implanted into human heart are used to treat heart arrhythmia. The modern pacemaker is an intellectual device that analyzes the patient condition at the lowest consumption of electric power. The pacemaker is implanted into human body inside a titanium capsule under the collarbone. The endocardiac electrode (ECE) of the pacemaker is drawn through the subclavian vein into the ventricle of the heart. The cathode tip ECE is pressed against the His-Purkinje nerve center. Due to current flow through the heart tissue, a double electrical layer (DEL) is formed near the cathode where charges of the blood electrolyte are separated. The equivalent scheme of the current flow is shown in figure 1.

The active resistance of the circuit - R_Σ is a sum of all series resistances of the human body R_{hb}, wires R_w, thin film on the electrode R_{tf} and a shunt R_s used in the current measurements. The Faraday resistance of DEL - R_f measured in experiment is approximately 10^3-10^4 Ohm and it is significantly larger than R_Σ. In this case, in conditions of direct current, the voltage drop on DEL is practically equal to the voltage of the power source.

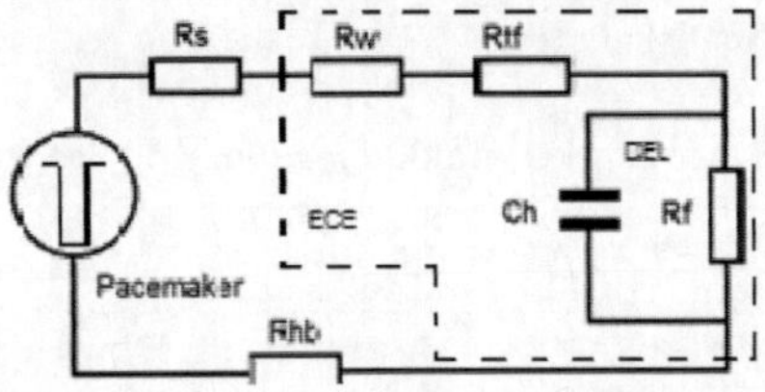

Figure 1. Schematic circuit of current flow from the pacemaker into a human body.

In the case of impulse voltage applied, electric capacitance of DEL must be taken into account. The electric capacitance of the cathode tip is much smaller than that of the anode tip ECE due to larger specific surface of the anode. During a short heart stimulating impulse (10^{-3} - 10^{-4} s), the drop of the voltage on the anode is much less than that on the cathode. Therefore, electrochemical processes near the anode can be neglected. Within a lapse time, the Helmholtz capacitance of DEL near the cathode, shorts Faraday resistance, and decreases ECE impedance significantly. Periodically repeated electric impulses of stimulation are mathematically represented in the form of super-position of harmonics with frequencies proportional to the heart frequency. For the harmonic with frequency ω, the impedance $Z(j\omega)$ of the equivalent current is: $Z(j\omega) = R_\Sigma + R_f/(1+j\omega\, R_fC_h)$.

At $\omega R_fC_h \to 0$, the impedance is maximum and is equal to $R_\Sigma + R_f$, and at $1/\omega R_fC_h \to 0$, it has minimal value equal to R_Σ. The above analysis shows that increase of the cathode capacitance C_h is a way to promote the charge transfer to the human body. Increase of the capacitance can be realized by increase of the specific surface of the electrode, which is in contact with the blood electrolyte. As it is shown in [1,3], vacuum PVD technologies of film deposition make it possible to produce the surface with highly developed structure of micro- and nano-sized pores (porous, columnar, fractal, etc.).

To coat tips of pacemaker electrodes, several PVD technologies were used: electron beam evaporation (EB PVD), magnetron sputtering (MS PVD), and evaporation in stationary and impulse arc (ARC PVD) produced in “Plasma Focus”.

Coatings were deposited on disc samples with diameter of 9 mm. Generally, titanium nitride was deposited. The thickness of PVD coating, measured in the cross-section, varied within 3-5 µm. In the case of ARC PVD, samples were preliminary sand blasted, then titanium powder was deposited in Ar plasma torch at atmospheric pressure. In dome cases, 80 keV titanium ion beam was used for finish treatment of the coating to produce nano-sized structure.

The coating properties in physiological solution were controlled in the electrochemical cell. The cell was powered by harmonic and impulse electric signals with the maximum voltage amplitude of 5 V. The cyclic volt-ampere characteristics (CVAC), the amplitude-frequency characteristics (AFC), and the transient characteristics were measured at pulsed voltage applied. In most cases, CVAC had a form of ellipse, the main axis of which determines the active reactive components of the current flowing through the cell.

Table 1 shows the results of the transition impedance change of AFC measurement of TiN samples, not coated–Z_1 and coated–Z_2 by ARC PVD technology. Table 2 shows the Helmholtz capacitance and the impedance of samples coated using different methods.

Though EB PVD technology demonstrated good capacitance, we did not use for production of pole pieces of pacemakers. We have not managed to obtain good adhesion and heterogeneity on the surface. Besides, it is difficult to coat very shapy surfaces.

Impulse ARC PVD seems can give good results, but this technology needs further research. Besides, the installations are rather complex, and the production must be expensive. There are no industrial impulse ARC PVD facilities yet.

MS PVD technology has many industrial applications, and it was used by W.C. Heraeus GmbH & Co. KG, Hanau, Germany for coating of pacemaker electrodes [3]. The drawback of this method is the high consumption of the target material. Only 30% of the target is used before its replacement, and not all sputtered material is deposited on the parts under deposition. This makes the production expensive in the case expensive materials (Pt, Ir) are used.

Table 1. Ratio of impedances of samples without coating – Z_1 and with ARC PVD coating – Z_2.

f, Hz	1	2	3	4	5	5	7	8	9	10
Z_1/Z_2	5.41	3.87	2.98	2.44	2.16	2.00	1.87	1.78	1.72	1.64

Table 2. Electrochemical properties of coatings obtained by different PVD technologies.

f, Hz	EB PVD TiN		MS PVD TiN		ARC PVD TiN		ARC PVD(impulse) "Plasma Focus"-Ti	
	C_h, μF	*Z*, Ohm	C_h, μF	*Z*, Ohm	C_h, μF	*Z*, Ohm	C_h, μF	*Z*, Ohm
2	20-25	2600-1600	7-20	1100-2400	7-18	850-1200	7-25	1600-2600
10	7-9	900-1100	2-6	700-2000	1.5-4	600-1000	3-9	700-1900
100	0.2-0.3	600-750	0.2-0.3	550-1100	0.15-0.25	570-950	0.5-1.3	500-1100

ARC PVD technology is more suitable for pacemaker electrodes production. In this case, TiN is synthesized in large volume of a vacuum chamber, the target material is used effectively (100% practically), and the production cost is low [4]. To make the adhesion better, the substrate was preliminary sand blasted to produce a coarse surface. Before TiN vacuum deposition, Ti powder coating of the thickness of 100 μm was deposited at atmospheric pressure from argon plasma torch. The roughness of such Ti layer is connected with the size of the powder particles. A 5 μm thick TIN coating with the columnar structure was deposited on this surface by ARC PVD technology. The last stage of the coating production is treatment of the surface to the depth ~ 200 nm by titanium ions with the energy of 50-80 keV. The biocompatible coating with large surface named "*Cardiocotin*" was produced by this method.

The "*Cardiocotin*" coating formed on tips of ECE are protected by [5,6]. Results of tests of ECE with "*Cardiocotin*" coating were compared with results of tests of ECE of other producers, and results are showed in figure 2. At the frequencies exceeding 5 Hz, the ratio of the capacitance current to the total current through ECE is the most high if "*Cardiocotin*" is used. Decrease of the performance for frequencies below 5 Hz is connected with formation of oxides on titanium during the anode operation cycle, which is known as the "valve effect".

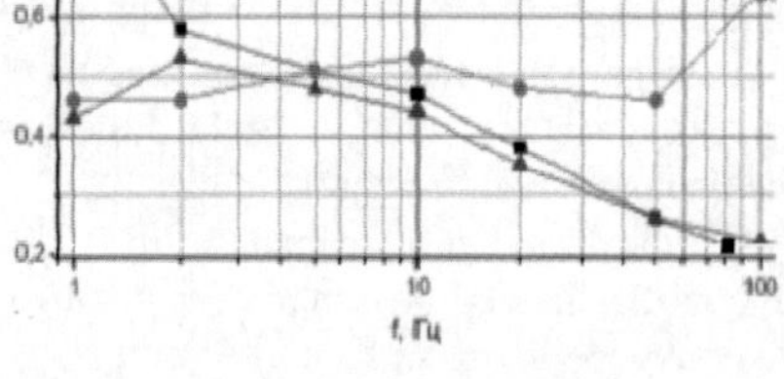

Figure 2. Ratio of ECE capacitance current to the total current as a function of frequency for ECE produced by "Cardiocotin" (red points) and ECE produced by St. Jude Medical Corp., USA (black) and Cardioelectronics Ltd., Russia (blue).

Clinic tests of electrodes with "*Cardiocotin*" coating were performed in 2007 and in 2013 and gave positive results. Figure 3 shows two basic parameters of ECE during clinic tests that determine ECE operation efficiency, namely, volt stimulation threshold (ST) of myocardium contraction stability and average for the ECE impedance impulse measured under GOST 31582-2012. The low level of ST shows stable work of biocompatible material of coating in the human body during six months. High level of impedance of ECE means high efficiency of transformation of energy and increase of the life time of pacemakers battery.

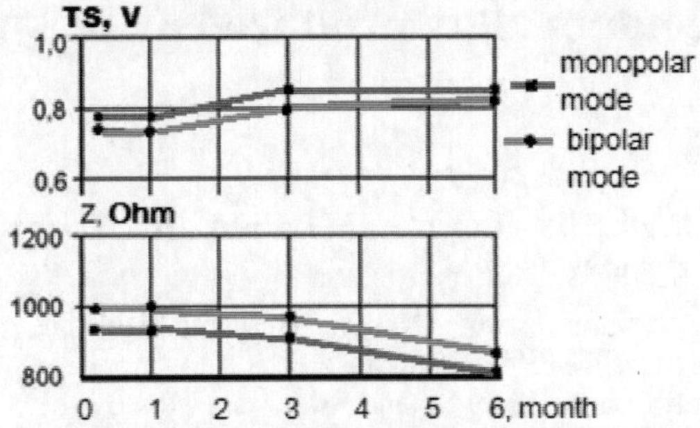

Figure 3. Results of control myocardium stimulation threshold (ST) and ECE impedance (Z) during clinic tests.

Based on clinic tests, "Cardiocotin", produced by Cardioelectronics Ltd, Klimovsk, is used in medical practice since 2013.

Acknowledgements

This work was supported in part by National Research Nuclear University MEPhI in the framework of the Russian Academic Excellence Project (contract No. 02.a03.21.0005, 27.08.2013).

References

[1] Schaldach M 1992 *Electrotherapy of the heart* (Springer-Verlag) 256 pp
[2] Barsukov E, Macdonald J R 2005 *Impedance Spectroscopy Theory, Experiment, and Applications* (John Wiley & Sons, Inc.) 595 pp
[3] Specht H, et al 2005 *Structural properties of PVD coatings on implants and their influence on stimulation performance in pacing applications* Proceeding of Material & for Medical Devices Conference Nov. 14-16 Boston 169-173
[4] Obrezkov O I, et al 2010 *Med. Tech.* **3** 13
[5] Obrezkov O I, et al 11/24/2014 Patent of invention "*Thin-Film Coating of Pole Pieces of Endo-cardiac Electrodes of Pacemakers and Method of Obtaining It*" RU 2538880 24.11.2014
[6] 01/24/2014 Certificate of trademark (service mark) "*Cardiocotin*" 504767 24.01.2014

Journal of Physics: Conference Series **748** (2016) 012019 doi:10.1088/1742-6596/748/1/012019

Stabilization of the high-temperature phases in ceramic coatings on zirconium alloy produced by plasma electrolytic oxidation

A V Apelfeld[1], S Y Betsofen[1], A M Borisov[1], B V Vladimirov[1], S V Savushkina[2] and E V Knyazev [3]

[1]Moscow Aviation Institute (National Research University), 4 Volokolamskoe highway, Moscow 125993, Russia
[2]Keldysh Research Center, 8 Onezhskaya st., Moscow 125438, Russia
[3]Intermix Met LLC, 2A Rabochaya st., Moscow 141400, Russia

Corresponding author e-mail : anatoly_borisov@mail.ru

Abstract. The composition and structure of ceramic coatings obtained on Zr-1%Nb alloy by plasma electrolytic oxidation (PEO) in aqueous electrolyte comprising 2 g/L KOH, 6 g/L $NaAlO_2$ and 2 g/L Na_2SiO_3 with addition of yttria nanopowder, have been studied. The PEO coatings of thickness ~20 μm were studied using scanning electron microscopy, X-ray microanalysis and X-ray phase analysis. Additives in the electrolyte of yttria nanopowder allowed stabilizing the high-temperature tetragonal and cubic zirconia in the coating.

1. Introduction

Plasma electrolytic oxidation (PEO) of zirconium alloys for obtaining ceramic coatings is the subject of many modern researches. For example, PEO of zirconium has been studied for biocompatible zirconium implants [1], for water corrosion protection of fuel elements of nuclear power reactors [2–4]. Due to low thermal conductivity and high thermal resistance, zirconium oxides are widely used as the top thermal barrier layers of thermal barrier coatings [5,6]. The possibilities of PEO for the thermal barrier coatings production were recently presented in [7-9]. Using zirconia PEO coatings at high temperatures requires stabilization of the high-temperature ZrO_2 phases. At transition of high-temperature tetragonal t-ZrO_2 to low-temperature monoclinic m-ZrO_2, the volume of material increases by 5-7%. It can lead to cracks and destruction of the coating. Stabilization of high-temperature phases is possible with additives of oxides CaO, MgO, Y_2O_3, Al_2O_3, etc. [10-11]. They form a solid solution with ZrO_2 and prevent phase transitions. The additives of these oxides allow retaining the metastable t-ZrO_2.

The aims of the present study were synthesis and investigations of oxide coatings on Zr-1%Nb alloy produced by plasma electrolytic oxidation in the electrolyte comprising an addition of yttria nanopowders. It is assumed that during PEO oxide nanoparticles can be incorporated into the structure of the coatings that will stabilize high-temperature phases of zirconia.

2. Experimental setup and characterization techniques

PEO coatings were formed on Zr-1% Nb alloy pipes with diameter of 8 mm. Specimens were ultrasonic degreased in acetone and ethanol, then rinsed in distilled water and dried on warm air. The coatings were formed by PEO in the electrolytes comprising 2 g/L KOH, 6 g/L $NaAlO_2$ and 2 g/L Na_2SiO_3 with addition of yttria nanopowder. Nanopowder particles were 10 ÷ 30 nm in diameter. Nanopowder was added to the electrolyte in amount of 2, 5, 8 and 10 g/L. The electrolytes were prepared in an ultrasonic bath with aim of uniform nanopowder distribution. PEO was carried out at an alternating current, and oxide coatings thickness of ~ 20 µm were obtained during 60 minutes of PEO with average current density of 10 A / dm^2.

During PEO, the current *I* flowing through the electrolytic bath and the voltage drop *U* on it were periodically measured. A typical *I-U* curve for one period of the alternating current with the frequency of 50 Hz is shown in figure 1. Rising branches 1 and 3 of *I-U* curve correspond to increase of the anodic and cathodic voltage, and falling branches 2 and 4 correspond to anodic and cathodic voltage decrease.

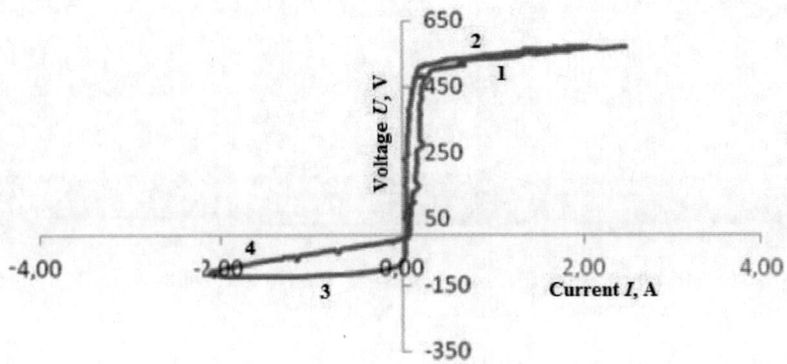

Figure 1. *I-U* curve that has been recorded after 1180 s of PEO.

It can be seen that the anodic voltage at first increases very rapidly with current increase that corresponds to ignition of the discharge, and then the *I-U* curve rises very slowly as in the case of normal glow gas discharge. At the large thickness of PEO coatings the rise of the *I-U* curve is also seen which is characteristic for anomalous glow discharge [12]. Transition of the discharge to the arc in the anode half-cycle of PEO is not observed. The difference of rising and falling branches of *I-U* curve indicates the formation of the anodic oxide film during the pulse of anodic voltage. Short-lived microdischarges moving along the metal surface in the anodic and cathodic half-cycles have significantly different character. The cathodic branches of the *I-U* curve correspond to the arc discharge. The voltage in cathodic half-cycles is not higher than 100 V. During PEO, the *I-U* curves do not undergo qualitative changes. The growth of the coating leads to increase of voltage of the discharge ignition. Detailed analysis of the evolution of the *I-U* curve revealed some changes for consecutive stages of PEO including anodizing, luminescence, sparking, and burning of microdischarges [13].

The structure of the coatings was studied using scanning electron microscopy (SEM), X-ray microanalysis and X-ray diffraction phase analysis (XRD).

3. Results and discussion

PEO coatings formed without yttria nanopowder additives are characterized by lenticular zirconium-enriched and silicium-enriched regions with fine globular structure (figure 2, a, b). In the center of some zirconia regions pores with a diameter of 20 µm are located. X-ray microanalysis shows that zirconium-enriched regions are composed of zirconia and silicium-enriched regions comprise SiO_2. The lenticular regions are typically composed of grains 0.2–0.4 µm in size (figure 2, c).

Journal of Physics: Conference Series **748** (2016) 012019 doi:10.1088/1742-6596/748/1/012019

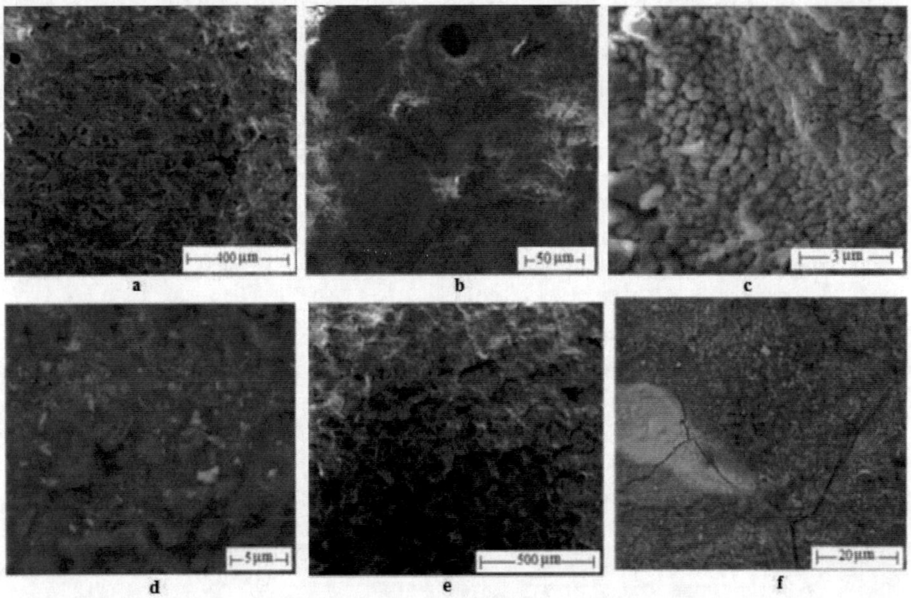

Figure 2. The SEM images of PEO coatings on Zr-1% Nb alloy without (a, b, c) and with additives of Y_2O_3 nanopowder of 5 g/L (d, e) and of 8 g/L (f).

If 2 g/L of yttria nanopowder is added in the electrolyte, the structure of PEO coating does not change significant, but if 5 g/L are added, conglomerates of yttria nanoparticles are observed on the surface (figure 2, d). X-ray microanalysis showed that the concentration of yttrium in the coating is higher than 3 at%. With 8 g/L additives of yttria in electrolyte the structure of PEO coatings became more rough (figure 2, e). Microcracks in the regions of zirconia and silica were observed (figure 2, f), also there were significant amount of conglomerates of yttria nanoparticles on the surface. X-ray microanalysis showed the content of yttrium in the coating was more than 8 at%.

X-ray diffraction analysis of the PEO coatings obtained without addition of yttria nanopowder showed monoclinic and tetragonal phases of zirconia in an equal amount (figure 3).

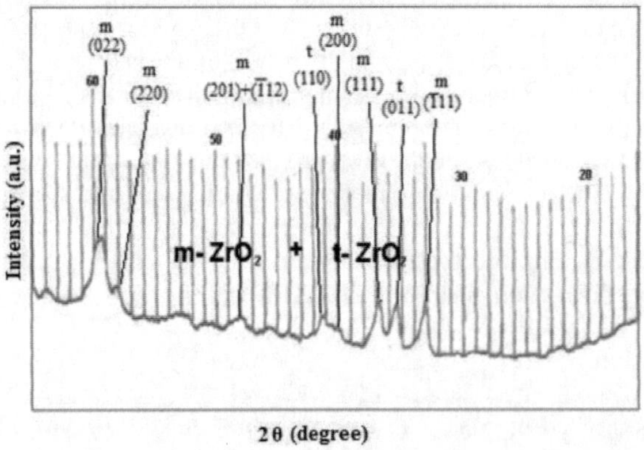

Figure 3. XRD pattern of the PEO coatings formed in electrolyte comprising 2 g/L KOH, 6 g/L $NaAlO_2$ and 2 g/L Na_2SiO_3 without additives of yttria.

If 2 g/L of Y_2O_3 nanopowder is added to the electrolyte, the ratio of monoclinic and tetragonal phases becomes 1: 4. If the concentration of Y_2O_3 nanopowder in the electrolyte rises to 5 g/L, only peaks of tetragonal phase are observed (figure 4, a). Finally, if the concentrations of Y_2O_3 nanopowder in electrolyte becomes 8 and 10 g/L, only peaks of cubic phase are observed in the XRD pattern (figure 4, b). Phase transformations obtained in this work correlate with the equilibrium diagram ZrO_2 - Y_2O_3. Increase of the yttria concentration leads to sequential formation of the monoclinic, tetragonal and cubic ZrO_2 phases. The transition monoclinic to tetragonal ZrO_2 phase takes place at 1170 °C. The cubic phase of zirconia is formed at 2370 °C [14]. The phase transitions to the tetragonal or cubic zirconia in PEO coatings can be explained by presence of yttria in the electrolyte in the form of a slurry. It can be assumed that yttria nano-particles decompose in microdischarges during the PEO process that results in dissolution of yttria in zirconia.

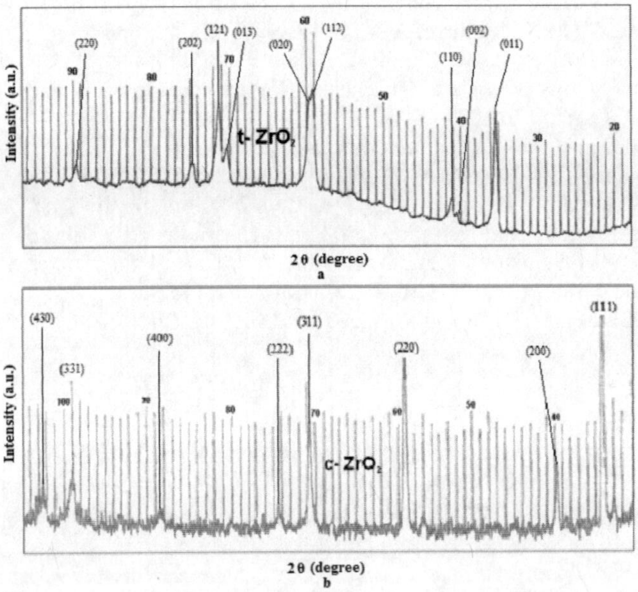

Figure 4. XRD patterns of the PEO coatings formed in the electrolyte comprising 2 g/L KOH, 6 g/L $NaAlO_2$ and 2 g/L Na_2SiO_3 with 5 g/L (a) and 8 g/L (b) additives of yttria nanopowder.

The presence of tetragonal zirconia in the PEO coating without yttria nanopowder in the electrolyte may be due to effect of stabilization of this phase by alumina due to presence of $NaAlO_2$ in the electrolyte. Stabilization of zirconia by alumina is known, for example, in the coatings obtained by magnetron sputtering [15].

4. Conclusions

The composition and the structure of ceramic coatings obtained on Zr-1%Nb alloy by plasma electrolytic oxidation in aqueous electrolyte, comprising 2 g/L KOH, 6 g/L $NaAlO_2$ and 2 g/L Na_2SiO_3 with addition of yttria nanopowder, have been studied.

It has been shown that additions of yttria nanopowder in the electrolyte allowed stabilizing the high-temperature tetragonal zirconia in the PEO coating. If 7 g/L of yttria nanopowder is added in the electrolyte, the high-temperature cubic phase of zirconia dominates in the PEO coatings.

References

[1] Legostaeva E V, Kulyashova K S, Komarova E G, Epple M, Sharkeev Y P and Khlusov I A 2013 *Mat.-wiss. u.Werkstofftech.* **44** 188
[2] Cheng Y, Matykina E, Arrabal R, Skeldon P, Thompson G E 2012 *Surf. Coat. Technol.* **206** 3230
[3] Borisov A M, Vostrikov V G, Ivanova S V, Kulikauskas V S, Lesnevsky L N, Lyakhovetsky M A, Romanovsky E A, Tkachenko N V, Tyurin V N 2013 *J. Surf. Invest.* **7** 437
[4] Borisov A M, Vostrikov V G, Ivanova S V, Lesnevsky L N, Lyakhovetsky M A, Romanovsky E A, Savushkina S V, Tkachenko N V, Tyurin V N 2014 *Inorg. Mater. Appl. Res.* **5** 245
[5] Slifka A J, Filla B J, Phelps J M , Bancke G and Berndt C C 1998 *J. Therm. Spr. Technol.* **7** 43
[6] Taylor R and Brandon J R 1992 *Surf. Coat. Technol.* **50** 141
[7] Borisov A M, Vostrikov V G, Romanovsky E A, Tkachenko N V, Vinogradov A V, Krit B L, Savushkina S V, Polyansky M N 2013 *J. Surf. Invest.* **7** 366
[8] Borisov A M, Savushkina S V, Vinogradov A V, Tkachenko N V, Vostrikov V G, Romanovsky E A, Polyansky M N, Ashmarin A A 2014 *J. Surf. Invest.* **8** 338
[9] Apelfeld A V, Borisov A M, Krit B L, Ludin V B, Polyansky M N, Romanovsky E A, Savushkina S V, Suminov I V, Tkachenko N V, Vinogradov A V and Vostrikov V G 2015 *Surf. Coat. Technol.* **269** 279
[10] Kelly J R and Denry I 2008 *Dent. materials* **24** 289
[11] Fabris S, Paxton A T and Finnis M W 2002 *Acta Materialia* **50** 5171
[12] Apelfeld A V, Lyudin V B, Dunkin O N, Nevskaya 2000 *Bull. Russ. Acad. Sci. Phys.* **64** 610
[13] Borisov A M, Efremov A P, Kuleshov E A, Krit B L, Lyudin V B, Mikheev V I, Polovtsev V A, Svetlakov V M, Suminov I V and Apelfeld A V 2002 *Bull. Russ. Acad. Sci. Phys.* **66** 1187
[14] Nakanishi N and Shigematsu T 1991 *Mater. Transact.* **32** 778
[15] Ji Z, Haynes J A, Ferber M K and Rigsbee J M 2001 *Surf. Coat. Technol.* **135** 109

Influence of soft X-ray of a vacuum spark with laser initiation on the surface properties of solid solutions $Cd_XHg_{1-X}Te$

O B Ananyin[1], G S Bogdanov[1], E D Vovchenko[1], I A Gerasimov[1], A P Melekhov[1], P S Krapiva[2], I K Novikov[1], R S Ramakoti[1], V G Sredin[1]

[1]National Research Nuclear University MEPhI (Moscow Engineering Physics Institute), 31 Kashirskoe highway, Moscow 115409, Russia
[2]All-Russia Research Institute of Automatics (VNIIA), 22 Sushchevskaya st., Moscow 127055, Russia

Corresponding author e-mail: apmelekhov@mephi.ru

Abstract. At a certain form of broadband source soft X-ray spectrum is expected to achieve selective radiation exposure to one of the elements of a multi-component material $Cd_xHg_{1-x}Te$. In this case we can talk about a change of the surface properties of the substance as a result of selective absorption of soft X-rays.

1. Introduction

It has been previously shown that the soft X-ray radiation from laser plasma (wavelength range $0.3 \div 30$ nm or energy range $0.1 \div 10$ keV, respectively) leads to a change in the surface morphology of the mono crystals and epitaxial layers of solid solutions $Cd_xHg_{1-x}Te$ [1,2]. In these works soft X-ray radiation of the laser plasma was used, which has over 80% of energy lying in the region of $50 \div 250$ eV. Thus there was no significant change in either carrier concentration or carrier of charge mobility in the bulk material and in the surface region. However, it can be expected that the increase in the degree of surface defects should lead to a change in its galvanomagnetic characteristics.

In this paper the effects on the semiconductor by the plasma generated by the laser-initiated vacuum spark hard X-rays with photon energies $E_{X\text{-ray}} = 0.5 \div 10$ keV are considered.

2. Experimental installation

The experimental setup consists of a vacuum interaction chamber (pressure $P \approx 10^{-5}$Torr), a high-voltage discharge system, initiating Nd: YAG laser ($\lambda = 1.06$ µm, the pulse energy $E_{laser} = 20 \div 35$ mJ and a pulse duration $\tau \approx 15$ ns) and diagnostic systems. In turn, the high-voltage discharge system consists of the conical electrodes, low inductance capacitor bank (capacitance $C = 0.011$ uF), a high vacuum current lead and high voltage DC ($U = 12 \div 15$ kV). The laser radiation is focused on the Fe-cathode. The anode is made of brass.

The vacuum spark forming region and a holder for mounting the irradiated sample are shown in figure 1. The distance from the discharge plasma to the sample is about 30 mm. To protect the sample from the optical range of the radiation and corpuscular streams Mylar filter with thickness about 3 µm is used with a 200 nm thick layer of Al deposited on it. The irradiation of the sample was done in the periodical pulsed mode with a repetition rate of discharge pulses $f \approx 1$ Hz.

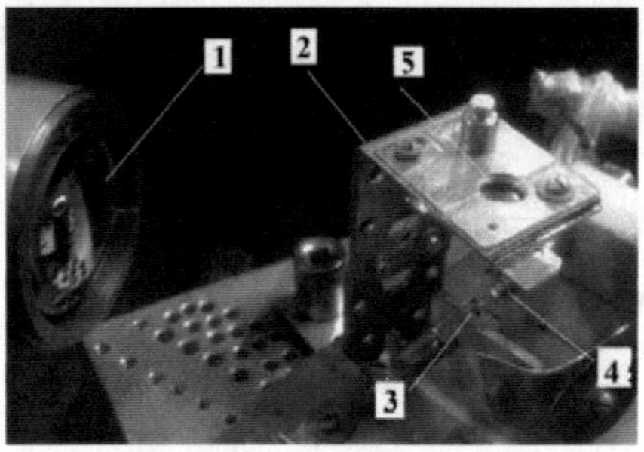

Figure 1. The area forming a vacuum spark in a pilot plant: 1 - lens for focusing the laser beam; 2 - bracket for mounting the sample; 3 - cathode (Fe); 4 - anode (brass); 5 - protective filter.

In order to control the charging current Rogowski coil with a time resolution better than 10 ns is used. Two identical *pin*-diode with a matched pair of edge filters were applied for the registration of X-ray radiation. They were placed inside the vacuum chamber at the distance of 20 cm from the discharge area.

3. Experiment and results

3.1. The spectrum of the source Soft X-ray from vacuum spark.

Experimental technique and method for reconstructing the X-ray spectrum are described in [3]. A form of the X-ray spectrum of a vacuum spark is shown in figure 2.

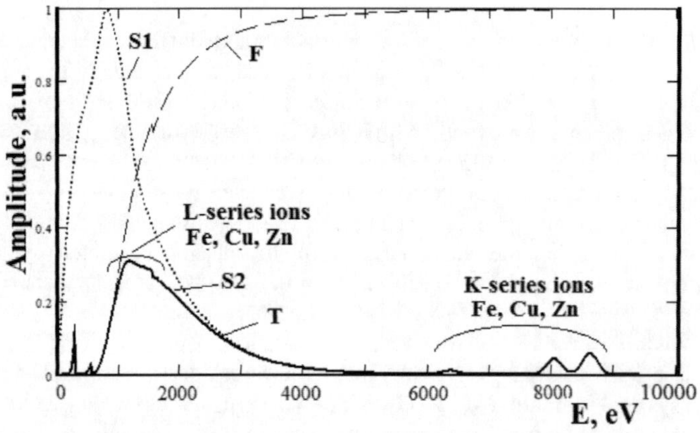

Figure 2. The figure shows the spectrum: 1) **S1** - synthesized by the experimental data range soft X-Ray induced laser-discharge plasma; 2) **S2** - spectrum soft X-Ray after Mylar filter; 3) **F** - transmission spectrum of 3μ Mylar filter; 4) **T** - *bremsstrahlung.*

The samples of semiconductor were irradiated by soft X-Ray from vacuum spark, with the above range. The latter is an epitaxial p-$Cd_{0.22}Hg_{0.78}Te$ and about 10 μm thick, grown in a coherent single-crystal substrate. The absorbed dose of soft X-Ray was estimated in the range of $0.15 \div 1.5$ J/cm^2, depending on the exposure time in the range of 15 ÷ 150 minutes. This estimate was obtained assuming that the soft X-ray source is a point source.

Figure 3 shows the spectra that characterize the spectral parameters of soft X-Ray source and model the X-ray transmission spectra of the investigated solid $Cd_{0.22}Hg_{0.78}Te$ solution [4].

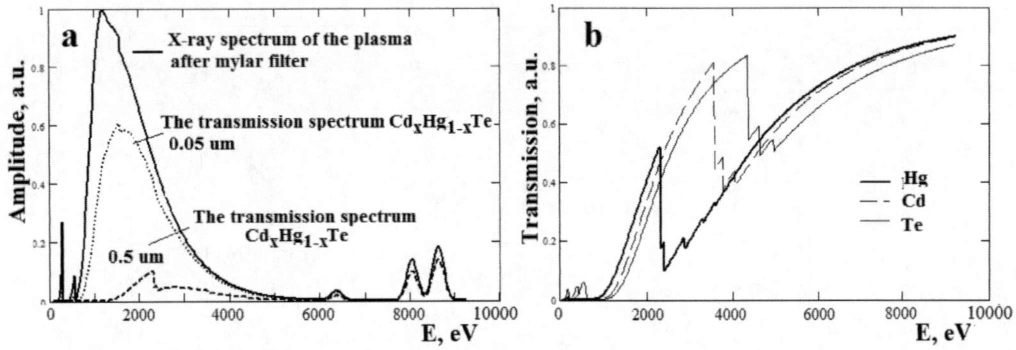

Figure 3. a - the calculated spectra of the plasma X-ray source and X-ray absorption in $Cd_xHg_{1-x}Te$ samples with the thickness of 0.05 μm (30%) and 0.5 μm (90%) [5]; b - the transmission of individual elements Hg, Cd, Te at the sample thickness of 0.5 mm [6]

Figure 3a shows the calculated spectral characteristics of the X-ray source for irradiating semiconductor and transmission of X-rays of the investigated solid solution $Cd_xHg_{1-x}Te$ with $x \approx 0.22$. One can see the presence of the transmission spectrum characteristics of the semiconductor in the range of 2.3-3.6 keV due to the presence of mercury in the sample. In figure 3b the transmission spectrum of mercury structure consisting of absorbing shocks at the following levels: MI (3.56 keV), MII (3.28 keV), MIII (2.85 keV), MIV (2.38 keV) and MV (2.29 keV) - is shown. Furthermore it should be understood that the intensity of radiation X-ray source used sharply decreases in the area of characteristic absorption of ions of Cd and Te (figure 2). From a comparison of figure 3a and figure 3b it follows that radiation of the X-ray source will be absorbed primarily by the Hg atoms, resulting in increased probability of the atoms migration from the lattice points and thus formation of the defects in the crystal structure [7]. This in turn can affect the electrical properties of the material.

3.2. Investigations of surface properties of irradiated samples p-$Cd_xHg_{1-x}Te$

Soft X-ray penetrates samples $Cd_xHg_{1-x}Te$, as shown above, to a depth of about 1 μm. To study the processes of its effect on the properties of the material it is necessary to use techniques that are sensitive to changes in the surface properties of semiconductors. The latter includes capacitive spectroscopy. Therefore, the samples were subjected to irradiation, conducted research capacity characteristics according to the method discussed in [8-10]. It was found that the temperature dependence of the capacity-voltage characteristics (CV characteristics) after exposure to X-ray radiation in the form of a singularity "shelf" due to the influence recharge of surface states (figure 4, *C* is the specific capacity of the MIS (metal- insulator- semiconductor) structure in the F/m^2). This dependence can be considered with good accuracy as a linear. From this it follows that the density of the surface states concentration of majority carriers within the thickness of the space charge is uniformly distributed. Furthermore, from the slope of this line integrated concentration of the majority carriers of charge can be determined.

Journal of Physics: Conference Series **748** (2016) 012020 doi:10.1088/1742-6596/748/1/012020

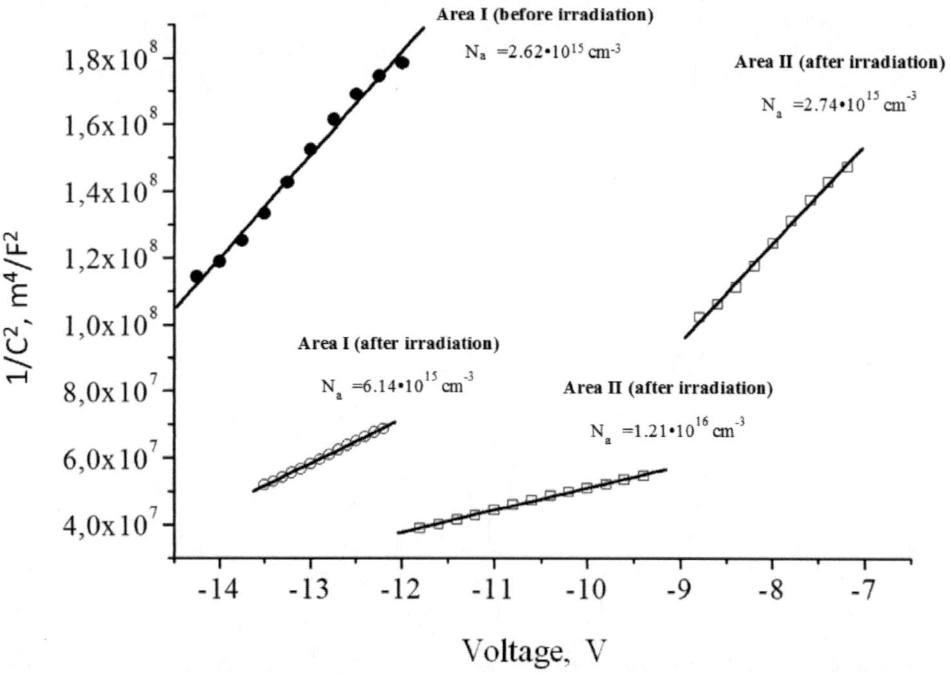

Figure 4. Dependence of 1/C2 in the depletion mode, the bias voltage to the source of MOS structures based on p-HgCdTe before and after the soft X-ray, built according to the capacitance-voltage measurements at a temperature of 770K at a frequency of 50 kHz with a direct scan voltage.

Processing of measurement results presented in figure 4 based on the assumptions above, showed that the concentration of holes in the surface layer to the source of MOS structure is $2.6 \cdot 10^{15}$ cm^{-3} and $2.7 \cdot 10^{15}$ cm^{-3}, and for MIS structures after exposure to X-ray radiation – $6.1 \cdot 10^{15}$ cm^{-3} and $1.2 \cdot 10^{16}$ cm^{-3}, respectively [11]. Thus, the concentration of majority carriers in the surface layer of the semiconductor after exposure to soft X-ray increases by 2-4 times, which may be related to the preferential knocking out of mercury ions and thereby forming defects of the acceptor type.

Further analysis of CV characteristics shows that original samples surface states are missing, while they appear in the irradiated samples. For one of them after exposure by soft X-ray maximum surface states appears at energies E_{v+} (0.37-0.45) eV (or E_{c-} (0.065-0.145) eV) at an average density of states in the energy range of $1.63 \cdot 10^{12}$ eV^{-1} · cm^2. In another area of the sample after exposure maximum of surface states appears at energies E_{v+} (0.34-0.39) eV (or E_{c-} (0.125-0.175) eV) at the average density of surface states in the energy range of $7.40 \cdot 10^{12}$ eV^{-1} · cm^2. Here, E_v and E_c are the top of the valence band and the conduction band bottom, respectively.

4. Conclusion

Thus, for the first time it has been shown that the use of a source of X-ray with the same spectrum as in Figure 2 for irradiation of solid solutions $Cd_{0.22}Hg_{0.78}Te$ creates a surface energy levels on the surface of epitaxial layers of this substance. The concentration of the majority carriers in the surface layer of the semiconductor after exposure to X-rays increases by 2-4 times. We note that our investigations were carried on samples excluding the crystallographic orientation of the surface, although earlier it was found that for these solid solutions such relationship exists. Therefore, it can be expected that the results of X-ray exposure will depend on this factor.

Acknowledgements
This work was supported in part by National Research Nuclear University MEPhI in the framework of the Russian Academic Excellence Project (contract No. 02.a03.21.0005, 27.08.2013).

References

[1] Sredin V G et. al. 2013 *Bulletin of the higher educational institutions Physics* **9** (2) 113
[2] Sredin V G et. al. 2013 *Applied physics* **6** 17
[3] Alkhimova M A et al. 2015 *Phys. Procedia* **71** 181
[4] Voytsehovskiy A V et al. 2008 *Physics and engineering of semiconductors* part 42 11
[5] http://henke.lbl.gov/optical_constants/filter2.html
[6] http://physics.nist.gov/PhysRefData/FFast/html/form.html
[7] Boltar K O et al. 2013 *Bulletin of the universities. Ser. Physics* part 56 8
[8] Sredin V G et al. 2015 *Bulletin of the higher educational institutions. Physics* t. 58
[9] Voytsehovskiy A V et al. 2005 *Bulletin of the universities. Ser. Physics* **6** 31
[10] Voytsehovskiy A V et al. 2009 *Bulletin of the universities. Ser. Physics* 10
[11] Koleshko V M, Kaplan G D 1977 *Reviews in electronic engineering* **2** (465) 82

Journal of Physics: Conference Series **748** (2016) 012021 doi:10.1088/1742-6596/748/1/012021

Application of DC plasma torch for synthesis of carbon nanostructured materials

M B Shavelkina, R H Amirov, V A Katarzhis and V I Kiselev

Joint Institute for High Temperatures of Russian Academy of Sciences, 13/2 Izhorskaya st., Moscow 125412, Russia

Corresponding author e-mail: mshavelkina@gmail.com

Abstract. The results of the synthesis of carbon nanostructures at high temperatures using a DC plasma torch are presented. Plasma was generated by introduction of argon, nitrogen and helium into the plasma torch with an anode in the form of an expanding channel. Sustainable modes of the plasma torch operation have been achieved by simultaneous tangential input of a plasma gas with a carbon source. Obtained solid products were studied using electron microscopy, thermogravimetry, Raman spectroscopy and X-ray diffraction to characterize their properties and morphological structures.

1. Introduction

Plasma is a useful tool to synthesize carbon nanomaterials, including diamonds, fullerenes, nanotubes, and graphene, as shown by Gonzalez-Aguilar et al [1]. High temperature of plasma synthesis and anhydrous conditions provide conditions for formation of particles with unique and desirable surface properties and morphology unreachable by other methods. These particle's properties are critical to their performance in application of materials.

2. Methods

For the synthesis of carbon nanostructured materials thermal plasma generator was used which is a high current divergent anode-channel DC plasma torch. A detailed description of the experimental setup was given in the studies [2]. The experiment involved a simultaneous input of hydrocarbons or soot with a plasma forming gas into the plasma torch, wherein heating and decompositions occurred in the plasma jet and in the region of the arc discharge, followed by condensation of the synthesis product on the metal surface of the target in a vacuum chamber, see figure 1. The carbon flow rate, the composition of the catalysts (for solid precursor), plasma gas flow and the power of the plasma torch varied independently to each other. For the experimental conditions the electric power of the plasma torch was set up to 40 kW. Helium, nitrogen and argon were used as plasma forming gases.

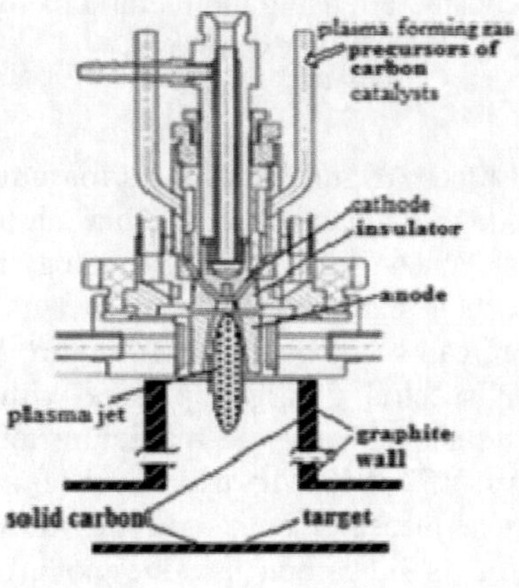

Figure 1. Schematic view of a plasma jet reactor.

For each experiment, the input of hydrocarbons was carried out after establishment of the temperature field in a graphite reactor (determined by the temperature of water in the cooling system) and establishment of the electrical characteristics of the plasma torch. The current value of the plasma torch was constant during the experiment and equaled to 350Å for argon, nitrogen and 400 Å for helium. Arc voltage was changed from 60 to 110 V depending on the gas pressure and the gas flow rate.

Soot, alcohols and hydrocarbons (methane, acetylene, and mixture of propane-butane in the ratio 30:70 mass %) were used as the source of carbon. Decomposition of soot in plasma jet reactor was performed in the presence of catalysts (Y_2O_3, Co, Ni and Fe). Their flow rate varied from 0.008 to 0.035 g/s. Experiments were conducted at various pressures and flow rates of hydrocarbons. Pressure changed from 150 to 720 Torr, and hydrocarbon flow rate varied from 0.05 to 0.37 g/s. Flow rates of alcohols reached 4 ml/min. Duration of the experiments was 5-20 min. The experimental conditions are presented in table 1.

Table 1. Technological conditions.

Power (kW)	Current (A)	Voltage (V)	Gas pressure (Torr)	Helium flow rate (g/s)	Nitrogen flow rate (g/s)	Argon flow rate (g/s)	Soot flow rate (g/s)	Hydro-carbons flow rate (g/s)
30-40	300 - 400	60-110	150-740	0.5-0.9	3.0-3.5	3.0-3.75	0.008-0.035	0.05-0.47

Method of electron microscopy was used to investigate the structure of the synthesized products on a scanning electron microscope of MIRA 3 TESCAN with Schottky field emission cathode in high vacuum regime. Raman spectra were investigated by using the exciting radiation at a wavelength of 532 nm (NTEGRA spectrum). X-ray diffraction data were obtained at the room temperature on a powder diffractometer Stoe Stadi P. To evaluate the specific surface area of materials the classic BET method was applied using a low-temperature nitrogen adsorption analyzer Quadrasorb SI.

Efficiency of the synthesis, thermal stability and phase composition of the carbon products were evaluated by thermogravimetry and differential scanning calorimetry on a synchronous thermal analyzer STA 409PC Luxx (NETZSCH) with the linear heating sample in air at the rate of 10 K/min at temperatures up to 1000°C.

The spectra of argon and helium plasma were taken on a three-channel fiber-optic spectrometer AvaSpec 2048 with the spectral resolution of 0.1 ÷ 0.4 nm and the spectral range 200-1100 nm.

Journal of Physics: Conference Series **748** (2016) 012021 doi:10.1088/1742-6596/748/1/012021

The electron temperature was considered, using the method of relative intensities of the same lines of ionization particles (the method of Boltzmann exhibitors).

3. Results and discussions

Design of the DC plasma torch to 40 kW (figure1) provided for simultaneous input of plasma gas and carbon precursor under angle. Voltage fluctuations in arc plasma torches of this type do not exceed 5%. At transition from one plasma forming gas (argon) to another (helium) and further addition to helium plasma of gaseous hydrocarbon (e.g. methane) there occurs a voltage increase and respective power increase, however, the arc remains stable (figure 2). Figure 3 shows the relative population of excited Ar states calculated from spectral of argon plasma with a liquid hydrocarbon (ethanol). Analysis of such experimental data showed that the maximum temperature of argon plasma at the energy input 40 kW and 350 A was 15700 K. The maximum temperature of argon in the stream of carbon particles was 16700 K. These measurements show that introduction of ethanol into a plasma flow does not cool it. Introduction of a solid carbon also does not affect plasma temperature.

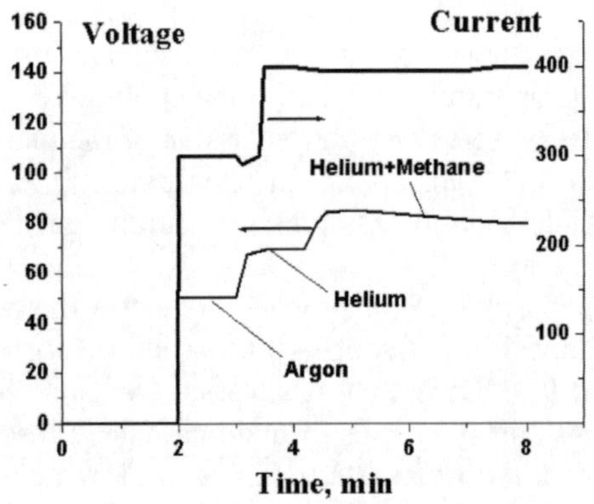

Figure 2. The effect of the stability of the plasma jet.

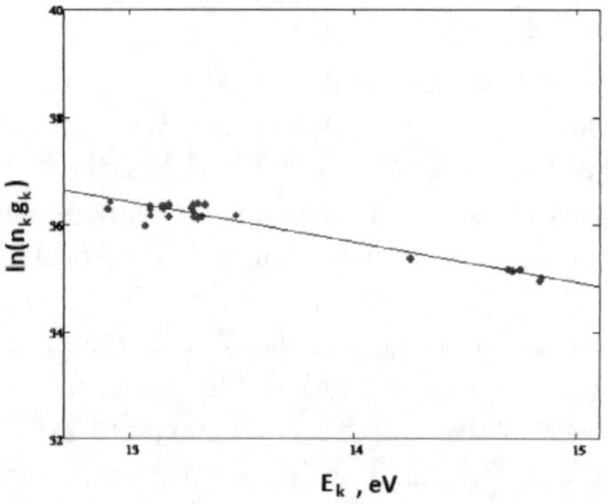

Figure 3. Relative population of excited Ar states.

Figures 4-5 presents the morphology of carbon nanostructured materials obtained by plasma jet.

Figure 4. SEM image of graphene sheets produced using helium plasma.

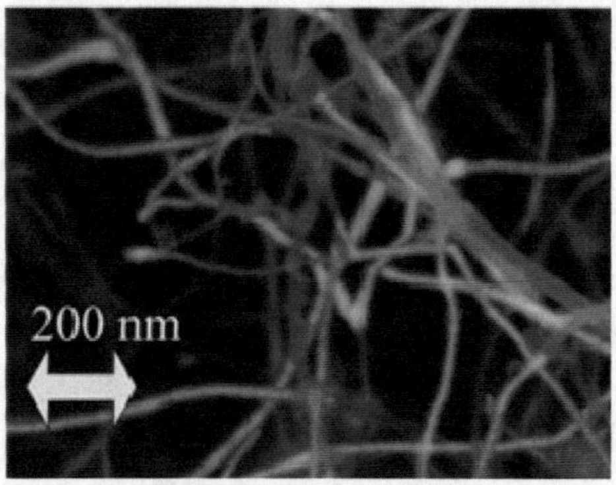

Figure 5. SEM image of carbon nanotubes produced using nitrogen plasma.

Journal of Physics: Conference Series **748** (2016) 012021 doi:10.1088/1742-6596/748/1/012021

Investigation of the synthesis products derived with the help of plasma jet has shown that by varying the medium pressure, plasma flow speed and rate of hydrocarbon materials input, it is possible to obtain various nanostructures of carbon – graphene (figure 4) as well as carbon nanotubes (figure 5). By varying the ratios of plasma gas and precursor in plasma jet it is possible to produce single-layer and multi-layer carbon nanotubes, in the shape of individual structures or collected in bunches, that is, bound nanostructures. The maximum production of carbon nanotubes was observed in helium plasma at combination $Ni/Co/Y_2O_3$ – 1.46/1.47/1.29 at.%, when the pressure was 500 Torr, arc current was 400 A and consumption of the mixture was 0.45 g/min. The diameter of the synthesized carbon nanotubes changed from 16 to 74 nanometers. Large graphene flakes having transverse lateral dimensions of 600 nm were synthesized at helium pressure of 350 Torr at the rate of 0.75 g/s. Distance between adjacent graphene flakes is 1-2 nm. Graphene materials may consist of one or 5 layers.

Typical bands were received on the Raman spectra of the investigated samples (figure 6) which are specific to carbon nanotubes and graphene – same as in the work [3].The sensitivity of the positions, widths and intensities of the D, G and 2D peaks–have made it possible to probe a variety of morphologies of graphene produced from different hydrocarbon materials decomposed by plasma jet.

X-ray diffraction patterns (figure 7) of carbon products synthesized from decomposition of methane in helium plasma of pressure of 350-710 Torr show rather narrow and sharp reflexes that are identified generally as the metal lines and hexagonal graphite.

Thermal oxidative behavior has shown the shift of oxidation temperature interval toward the higher temperatures area. It is believed that the obtained structures are single-phase and have a few defects. In some circumstances graphite appears as a second phase. The carbon nanostructures have different thermal stability formed by the pyrolysis of the carbon black with the catalysts and hydrocarbons decomposition. Less stable are graphene materials derived from methane (curve 1) and mixture of propane and butane (curve 2) (figure 8).

Specific surface area of the samples with the lateral size of 300-600 nm was 350-400 m^2/g measured using the BET method. Figure 9 presents the comparison of pore size distribution for two gas pressures used in the experiment. The main range of pore radii was in the region of 10-70 Å, corresponding to mesopores.

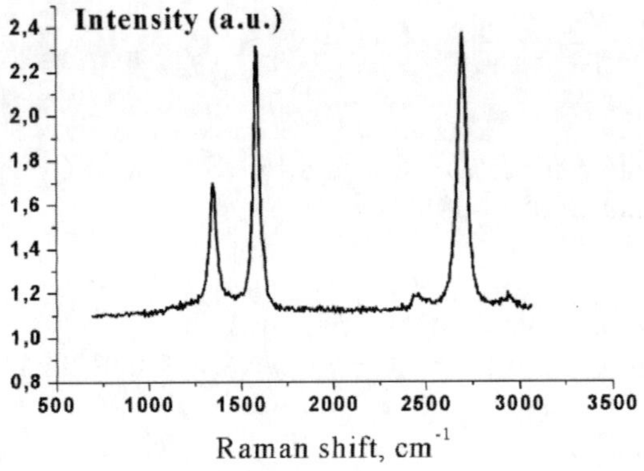

Figure 6. A typical Raman spectrum of few-layered graphene obtained from decomposition of methane by helium plasma. The three intense features are the D band at 1334 cm^{-1}, the G band at 1580 cm^{-1} and the 2D band at 2660 cm^{-1}.

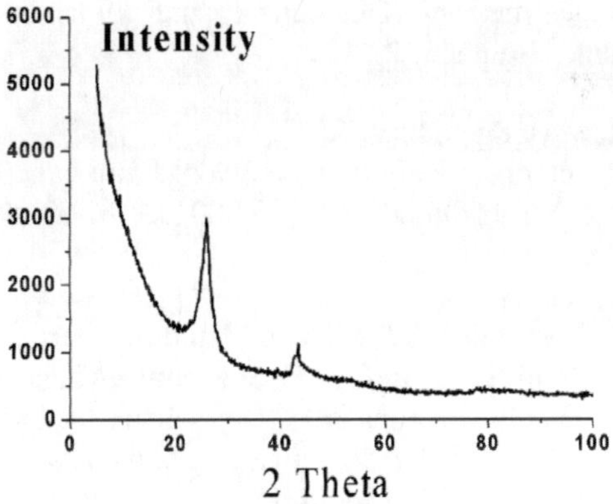

Figure 7. XRD pattern of graphene synthesized from decomposition of methane by helium plasma.

Thus, we have developed a method for synthesizing carbon nanostructured materials by plasma jet. The resulting few-layered graphene sheets and carbon nanotubes have fewer defects than those generated using chemical methods [4].

4. Conclusion

The efficiency of DC plasma torches with divergent channel anode, capacity of up to 40 kW for the synthesis of nanostructured carbon materials was demonstrated. Changing geometry of a reactor, pressure and velocity of plasma jet it is possible to vary the rate of carbon vapor cooling within a wide range. By varying the rate of carbon vapor cooling it is possible to obtain a wide range of carbon nanostructures with high yield of pure deposit.

On the basis of the research of the synthesis products conducted with the use of electron microscopy, thermogravimetry and X-ray analysis, the optimal conditions for the synthesis of each type of material: carbon nanotubes, carbon nanofibers and graphene were determined. Graphene structures synthesized using plasma jet are active for new catalysts.

As a whole, the experimental data allow step by step to scale the process of production of graphene and carbon nanotubes of desirable morphology.

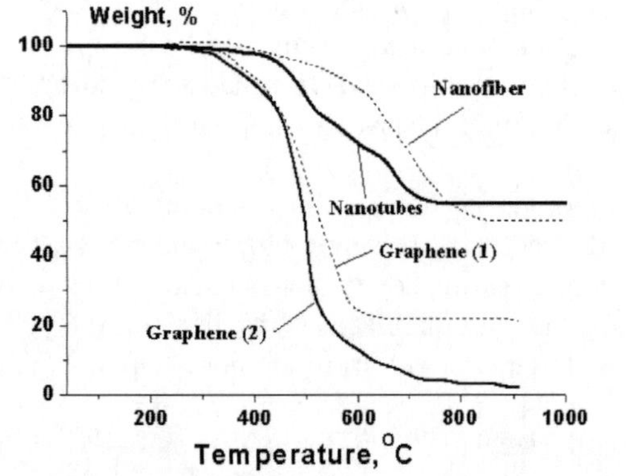

Figure 8. Comparison of the thermal stability of plasma jet synthesis products from various precursors: solid (nanofibers and nanotubes), gaseous methane (Graphene 1) and propane-butane (Graphene 2).

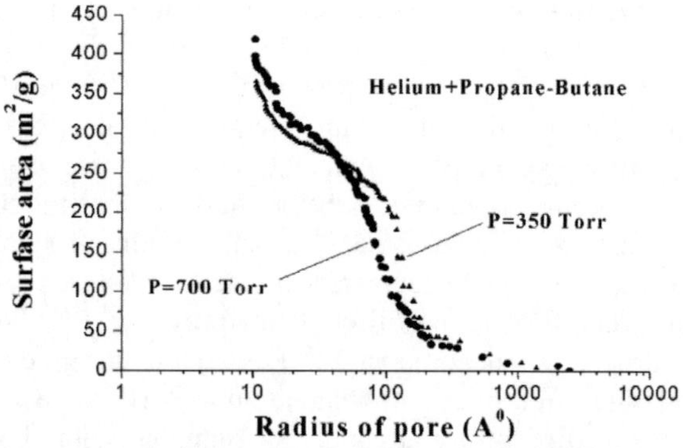

Figure 9. Specific surface area as a function of the pore radius.

Acknowledgements

The authors gratefully acknowledge Russian Foundation for Basic Research for the support by the grants No. 15-08-00165, 16-08-00145, 16-08-00057 and 16-08-00081.

References

[1] Gonzalez-A J, Moreno M and Fulcheri L 2007 *J. of Phys. D: Appl. Phys.* **40** 2361
[2] Amirov R, Isakaev E, Shavelkina M and Shatalova T 2014 *J. of Phys.: Conf. Ser.* **550** 012023
[3] Saito R, Hofmann M, Dresselhaus G, Jorio A and Dresselhaus M S 2011 *Adv. in Phys.* **60** 413
[4] Prasek J, Drbohlavova J, Chomoucka J, Hubalek J, Jasek O, Adam V and Kizek R 2011 *J. Mater. Chem.* **21** 15872